NATURAL LIGHTING

SolarVision Publications
Church Hill, Harrisville, N.H. 03450

Published by
SolarVision Publications
Church Hill
Harrisville, NH 03450

Director of Publications: William Dunkerley
Editor: Eileen Haas

Cover illustration courtesy of the Glass Division of the Ford Motor Company.

The publisher wishes to acknowledge and thank Donald W. Aitken, Harvey Bryan, Robert Clear, J.W. Griffith, James A. Moore, James Rosen, Stephen Selkowitz, and Marguerite Villecco for their contributions to this book, some of which were developed by the staff of *Solar Age* magazine, William D'Allessandro, Editor.

Library of Congress Catalog Number: 82-61224
Main entry under title:
Natural Lighting
Includes Index
ISBN 0-918984-05-X

Contents

Preface

In the days before electric lights, people did their daily work in front of windows, depending on the sky and the sun to give them enough light to do even the most critical tasks. Today we don't think twice about turning on the lights, even in a bright room on a sunny day.

The purpose of this book is to show how to save thousands of dollars in utility costs by opening up buildings to natural light. The cost-effectiveness of natural lighting has been more than proved, yet most architects never even take it into account when designing or renovating a structure.

Natural lighting means much more than just large windows. It includes toplighting and sidelighting techniques, sun shades and controls, and the use of reflective surfaces to beam light from a single source to several surrounding areas. It involves a whole new type of building design and unique calculations. This book delves into *all* the elements of daylighting design, including state-of-the-art techniques for using natural light.

We think it will prove invaluable to architects, engineers, and anyone else who wants to design for an energy-conscious future.

The Importance of Daylighting In Buildings

Chapter One

Lighting accounts for over five percent of total energy consumption in the U.S.

Lighting accounts for about 20 percent of total electrical energy consumption in the United States, or 420 billion kWh per year. This represents over 5 percent of total national energy consumption and is approximately equivalent, in terms of daily energy consumption, to the total output of the Alaskan oil pipeline.

The winter of 1973-74 saw many lights extinguished in response to an energy shortage. Energy conservation in lighting became associated with delamping, which in turn was viewed as "doing without" and a sacrifice in the quality of living or working environments. Energy conservation practices, however, can provide equivalent or improved visual performance and visual comfort while producing substantial energy and power savings. Four different elements in this process can be identified: 1) substitution of efficient lighting systems and components for less efficient products; 2) improved lighting design practice that eliminates wasteful energy use; 3) improved operation maintenance of lighting systems; and 4) a return to a partial reliance on natural lighting techniques.

Prior to the introduction of the electric light, the best light available in buildings for visual tasks was daylight. Candles, kerosene lamps, and gas lamps must have run a distant second to daylight as a source for indoor illumination. Incandescent lamps represented a major improvement over open-flame sources but were not capable of achieving illumination equivalent to daylight efficacy.

There has been little interest in daylighting in the U.S. for the last twenty years

At about the time that incandescents gave way to fluorescents in commercial buildings, other trends, such as rising urban land costs, the advent of building air conditioning systems, and low-cost electricity combined to eliminate daylight as an essential element in building design. As a result, there has been little serious interest or activity in daylighting in the United States during the last 20 years.

International events and the resultant energy crisis in 1973 forced a reexamination of energy use patterns in the building sector. Early energy conservation checklists advocated "increased daylight utilization" without indicating how, or in what circumstances, one could use more daylight to achieve significant savings. The same checklists, without concern for apparent contradictions, typically advocated reduction or elimination of window area to conserve heat. Five years later, one finds an increased awareness and interest in daylighting, but little that indicates daylighting design has yet found a comfortable niche in professional practices.

As architects and engineers focus on thermal load reductions, lighting looms as the single largest energy consumer in the building. A renewed interest in daylighting has generated claims of very substantial energy savings in newly designed energy-conserving office buildings. Although the potential exists, it is not certain that projected savings can be fully realized. Skills in daylighting design analysis have been rediscovered after 20 years of dormancy, but the experience which must temper knowledge to produce effective results is frequently lacking. In this book we will explore the fundamental components of good daylighting design and address the issues of how they fit in with other energy-conserving measures to produce truly energy-efficient structures.

The commercial sector

Natural lighting serves several important functions in buildings

Natural lighting serves several important functions in buildings. Architects have long recognized the visual power of a shaft of sunlight penetrating a dark church sanctuary or the beauty of a sunset lighting up a stained glass window. Our concern here is for more pragmatic use of natural lighting—to offset electrical lighting requirements in commercial and industrial buildings. The primary focus in the commercial and industrial sector is office buildings, schools, commercial low-rise structures, and warehouses. These building types are characterized by daytime use patterns, long hours of lighting use, relatively high lighting levels, and high installed watts per square foot. Lighting is thus a significant energy consumption factor in most of these building types and represents a large fraction of total building utility costs.

The range of illumination in which the human eye can function is a very broad one

Use of solar energy for thermal control in buildings has acquainted us with the sun as a source of Btu's. If we turn to the sun as a source of illumination, our perspective must change. While thermal comfort spans a relatively small range, perhaps 65° to 80°F, the range of illumination in which the human eye can effectively function is much broader.

A person with very good eyesight can read effectively with one or two footcandles of light. Typical offices are lit to 50 to 100 footcandles. Direct sunlight is about 10,000 footcandles. The efficacy of sunlight, or the number of lumens of light delivered per watt of energy, varies from about 90 lumens per watt for direct sunlight to as much as 150 lumens per watt for light coming from a deep blue sky. As a rough design figure, 100 lumens per watt is a good place to start. This can be compared to the efficiency of a variety of electric light sources. A fair approximation is that daylight is as efficient as the best electric lighting systems in its ability to deliver illumination per unit heat delivered.

Daylight is as efficient as the best electric lighting systems in its ability to deliver illumination

Economics

If electric lighting can provide good visual performance at very low energy use, the fraction saved by daylighting will be proportionately less than if daylighting is competing against a very inefficient electric lighting system.

In 1970, it was acceptable to design lighting systems for 100 to 200 footcandles in office buildings with electricity consumption of about 5 watts per square foot. By 1980, recommended light levels dropped down to a range of 25-100 footcandles, with typical power consumption dropping to perhaps 2.5 watts per square foot. Over the next few years, average footcandle levels are expected to drop somewhat further and, together with the use of more efficient lighting hardware, electricity consumption of lighting systems are expected to drop to 1.5 watts per square foot. These figures are for standard design. With state-of-the-art hardware and refined design, good lighting should be obtainable with the expenditure of no more than .5 to 1 watt per square foot.

In addition to energy savings, at least four other reasons favor consideration of natural lighting in buildings:

A daylit building in which the lights are dimmed should lower building peak demand

1. Commercial building owners and operators pay not only for the energy they consume but also for a demand charge, which reflects their peak power consumption each month. Since many utilities have difficulty in siting and building new power plants, there is tremendous pressure to restrict growth of new demand. Demand charges encourage building owners to make modifications in their buildings so that their loads are leveled. Since peak building loads frequently occur in the afternoon when both air conditioning and lighting are on, a daylit building in which the lights are dimmed or turned off should reduce building peak demand, and thus have an additional economic benefit for the owners.

2. The provision of natural light in a space makes the occupant less dependent upon mechanical systems in the building. Compare an office worker in a room with no windows, and totally dependent upon electric lighting and mechanical ventilation, to a worker in a perimeter office with operable windows and daylight. In the case of a power failure, the worker in the interior office would have to quit working and leave the building. The value of a worker's productivity for one hour per year is approximately equal to the annual energy costs of lighting in the space that the worker occupies. Thus the ability to continue work for the single hour per year in a daylit office would equal the entire potential energy savings that daylighting might provide in offsetting electric lighting.

Reducing the use of electric energy for lighting provides a degree of inflation protection

3. Every indication is that electric costs will continue to rise at rates equal to or exceeding the general rate of inflation. Reducing the use of electric energy for lighting provides a degree of inflation protection.

The qualitative aspects of daylight are generally pleasing to building occupants

4. The illumination on a horizontal surface expressed in footcandles is the generally used measure of lighting design. In fact, this measure tells us very little about the ability to see or perform a visual task, and lighting professionals have developed other metrics that more nearly measure visual performance rather than the illumination at a point. Studies have shown that one footcandle from a window can provide visibility equivalent to three footcandles from overhead. Other qualitative aspects of daylight, such as its color rendition and modeling effects, are generally pleasing to building occupants.

Views and visual connections with the outdoors are related benefits that accompany most daylit spaces.

Daylight and sunlight

Natural lighting techniques encompass the use of diffuse light from the sky, or daylight, and beam radiation from the sun, or sunlight. In addition, side lighting techniques, or the use of natural light through windows, and top lighting, or the use of skylights in buildings, are considered (**Fig. 1**).

Natural lighting techniques encompass the use of diffuse light and beam radiation

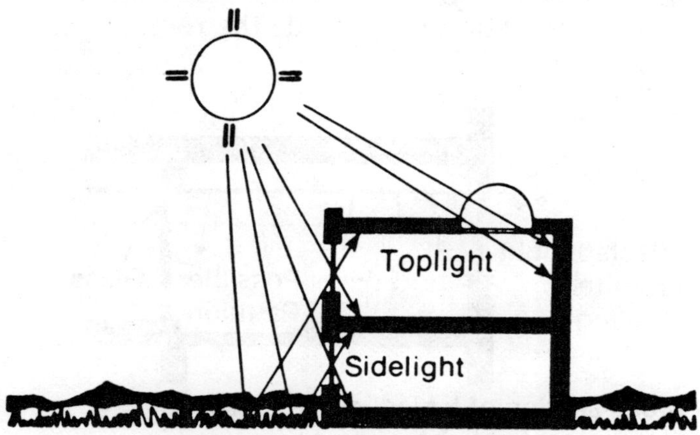

Fig. 1 — Lighting from windows and skylights.

Side lighting through windows typically utilizes diffuse radiation only. Direct solar gain, although occasionally pleasant, typically leads to overheating and thermal discomfort. Daylighting levels from windows in one wall of a room fall off rapidly as we move deeper into the room, away from the window wall. A typical practical limit for daylight penetration into an office is 15 to 20 ft. from the window. Some techniques are available for extending the depth of this perimeter zone.

Direct solar gain typically leads to overheating and thermal discomfort

Glass blocks have been used extensively to direct sunlight

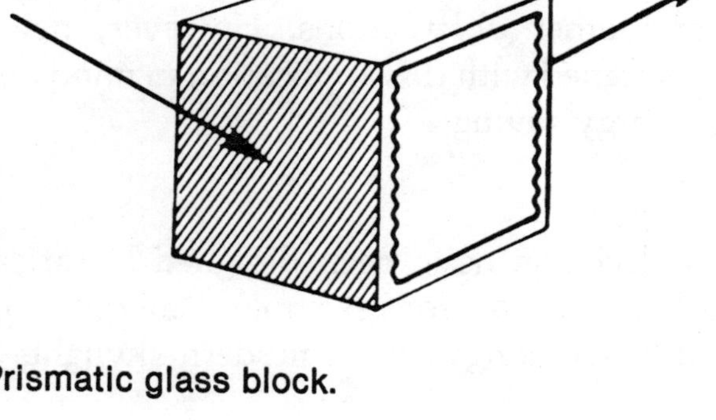

Fig. 2 — Prismatic glass block.

Glass blocks have been used extensively to direct sunlight deeper into rooms

deeper into rooms to complement diffuse light near windows (**Fig. 2**). Glass blocks often have ribs that provide some degree of light control for daily and seasonal variations in solar elevation and azimuth. Even deeper light penetration can be achieved by controlling sunlight directly. One concept, called "beam sunlighting," involves reflecting direct rays from the sun from silvered reflective venetian blinds mounted in the upper 2 feet of a typical window. The reflected rays are aimed toward the ceiling of the room to a maximum depth of approximately 30 to 40 feet (**Fig. 3**). The ceiling then acts as a diffuse reflector, providing diffuse illumination deep inside the room.

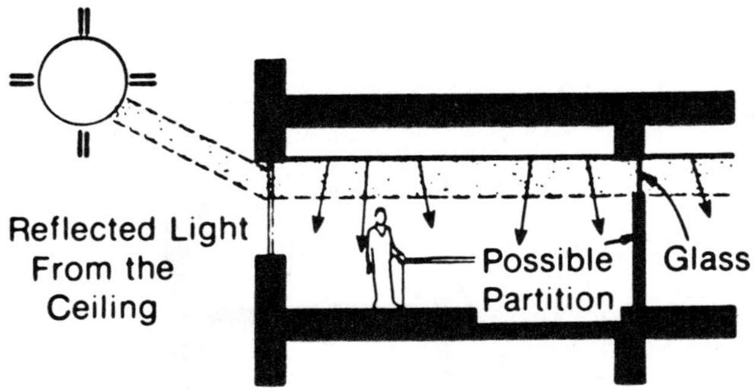

Fig. 3 — Cross section of typical office space (south perimeter location) showing position and operation of beam daylighting apparatus.

More sophisticated lighting controls are required for partly cloudy sky conditions

Good lighting quality is achieved by such a scheme, but the control of reflected light as sun angles change is a problem. A variety of controllable reflecting or refractor-type devices have been examined, but it is hard to maintain simplicity and low cost in these devices without sacrificing potential performance. In addition, more sophisticated lighting controls are required for partly cloudy sky conditions, in which case the sun's intensity will change sharply over very short time intervals. This concept appears to have only limited applications in existing buildings because of window and ceiling design characteristics and problems with shading from adjacent buildings and other obstructions. However, new buildings specifically designed with this application in mind might realize substantial energy savings.

Skylights

Although skylights have been designed historically to exclude direct radiation (for example, sawtooth designs or skylights with overhangs), most modern skylights lay flat in the roof plane and thus accept both direct and diffuse radiation. Light distribution within the space is significantly more uniform than with side-lighting applications. Although conventional skylights are limited to providing daylighting in one-story spaces or multistory atrium spaces, several concepts have been developed that use direct sunlight accepted through an aperture on the roof and then distributed throughout the building with a series of mirrors, lenses, and other optical controls. Although such schemes are technically feasible, their optical performance requirements and complexity make their use impractical.

The potential energy savings are significant: however, effective daylighting design requires that time and effort be expended to solve a series of problems and issues which currently act as obstacles to the widespread use of daylighting. There is a real danger in believing that good daylighting practice simply means the use of large windows and that the use of large windows guarantees significant energy savings. Four major points must be addressed before daylighting practice can be widely implemented in this country: 1) analysis and design techniques, 2) thermal/illumination tradeoffs, 3) sun and glare control, and 4) lighting controls. In the following chapters we will discuss each of them in turn.

Good daylighting practice does not simply mean the use of large windows

Design Methods And Strategies

Chapter Two

Analysis and design

Ask a building designer today how to design a room to provide 50 footcandles on a desk throughout 80% of the working hours of the year using daylight, and you are likely to get puzzled looks and quick shuffling through textbooks and lighting handbooks. Simply stated, there is a lack of effective, widely used design methods in the United States today. The problem is not a lack of design methods; 58 are listed in a 1970 CIE publication. Most, however, were originated in European countries where cloudy skies are the typical minimum brightness conditions. In much of the United States, clear sky conditions prevail, and many of the techniques developed for analysis of overcast skies are not suitable for use with the variable luminance distributions one finds under clear sky conditions.

There is a lack of effective, widely used design methods in the U.S. today

A variety of different design methods have been and continue to be used in this country. These include computational, graphical, tabular, nomographical, protractored, diagrammatic, (such as sky vault projections), and physical models. Very powerful computer models are available to compute footcandle levels and equivalent sphere illumination levels in rooms, but they suffer from an inability to easily model detailed effects inside a room, such as furniture placement. Although current programs are relatively costly, it appears likely that computer run time can be substantially compressed with significant cost reductions. Physical modeling techniques which are already widely used by architects for other purposes can provide a very versatile simulation of detailed conditions to be expected in a typical building.

Physical modeling can provide a very versatile simulation of conditions to be expected in a typical building

A more fundamental problem than the lack of design methods is the lack of awareness and knowledge of these design methods by practicing professionals and the schools now training the design professionals of the future. The lack of educational programs in the daylighting design field over the last 20 years has resulted in a generation of practicing architects and engineers who now have little academic or practical experience in daylighting design. This lack of professional design experience is compounded by the lack of well-documented examples of buildings incorporating effective daylighting. One can find many examples of architectural design with effective lighting, HVAC, and solar heating systems, but few, if any, which focus on effective daylighting solutions to building design problems.

Daylight availability

Having selected one of several daylighting design methods, the designer must confront the lack of information regarding daylighting availability in the United States. For a given location, are the skies characterized by clear, cloudy, or partly cloudy conditions? For what fraction of the working hours of the year can one expect certain minimum sky conditions to be exceeded? Data of this type exist for many European cities but for very few locations in the United States (**Fig. 4**). Measurements which were made some time ago in Washington, D.C. and other urban areas are now suspect due to atmospheric changes resulting from urban air pollution and changes in other climatic variables. Rather than wait several years before such new data can be collected, it may be feasible to generate daylighting availability data from the solar radiation data now being collected throughout the United States.

Measurements which were made some time ago in Washington, D.C. and other urban areas are now suspect because of air pollution

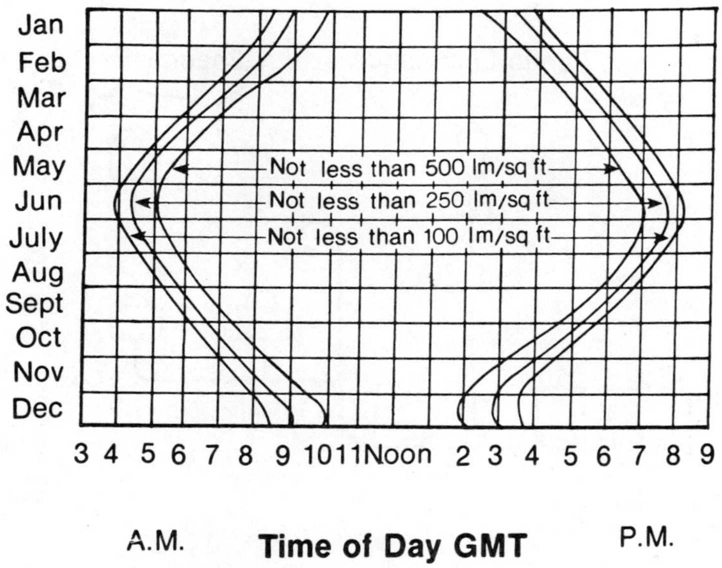

Fig. 4 — Average illumination on horizontal surface at Teddington, U.K., 51° N, 1929–33.

If you collect, at 15-minute intervals, total horizontal radiation; diffuse horizontal radiation; illumination received on four vertical surfaces facing north, south, east, and west; and data from a sky luminance sensor which records sky brightness directly overhead (**Fig. 5**), you can develop or select and then validate a computational procedure for converting the solar radiation data base into an illumination data base.

Building designs that have been optimized for daylighting use will have an impact on thermal gains and losses as well. Just as building design decisions that have been made to optimize thermal performance will inadvertently impact illumination issues, the sizing and placement of windows to maximize daylighting benefits will have thermal impacts which must be considered. A number of large computer programs are now available which will provide an annual energy analysis for large buildings. These are relatively complex programs which model building performance hour by hour throughout the year. Daylighting performance has been incorporated into several of these models in a limited way.

Building daylighting designs will have an impact on thermal gains and losses

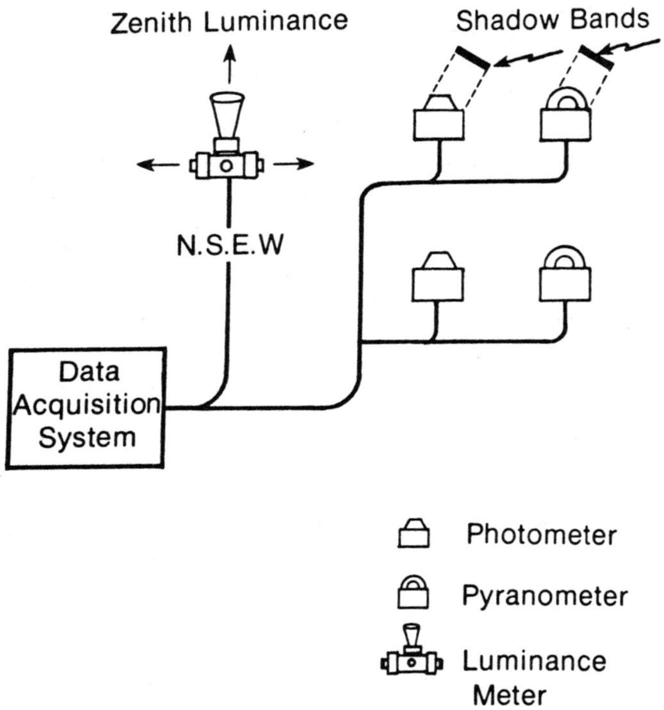

Fig. 5 — Instrumentation for solar illumination correlation study.

A similar but simplified approach is available for predicting the annual performance of skylights in buildings. In almost all cases, results indicate that consideration of daylighting benefits alters the determination of optimum window size toward larger windows or skylights than one would predict from a thermal analysis perspective alone.

Strategies of daylight design

Good daylight design does not simply mean large windows. It must be approached both quantitatively and qualitatively on broader and more sensitive design terms.

Daylighting seeks optimum amounts and areas of natural illumination for biological and task needs

Daylighting seeks optimum amounts and areas of natural illumination for biological and task needs, which means good quantity and distribution of natural light. The impact of heating and cooling must be addressed, and tradeoffs must be made appropriate to location and time.

Glare needs to be controlled to ensure that light will be comfortable and pleasant, as well as adequate in quantity. Appropriate surfaces need to be illuminated to dispel the perception of gloom.

Daylight is a dynamic phenomenon. Its relationship to natural rhythms of the surroundings requires a dynamic design response, specific to time and place. Designers must learn to approach the design process with this dynamic in mind and understand that the building will be visually and functionally different according to day and season. Daylight is not something one adds to building design; it is implicit to every design decision.

Daylight is dynamic; it responds to the natural rhythms of the surroundings

Factors affecting daylight design include variations in the amount and source of daylight, caused by the position and intensity of sunlight. The luminance and luminance distribution of clear, partly cloudy, and overcast skies must also be considered. The effects of local terrain, landscaping, and nearby buildings and the glare and luminance patterns within view are important, as is the color of daylight as combined with other sources of light (**Figs. 6 and 7**).

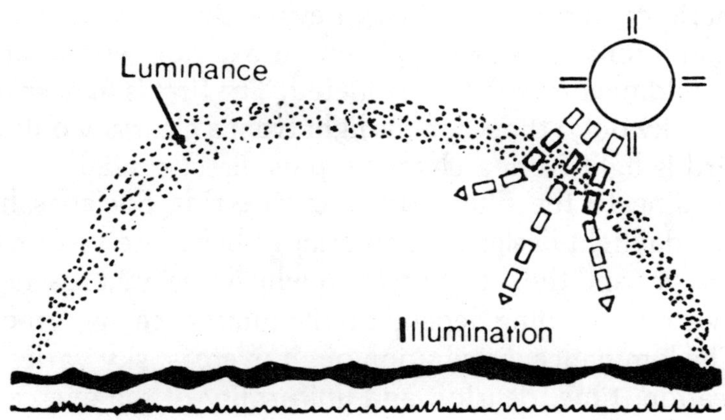

Fig. 6 — Visible radiant energy (lumens).

The design strategies used for admitting natural light into the building must be responsive to all of these factors. Natural light is a diffuse form of sunlight; daylight design must deal not only with direct solar radiation but with sky light and reflected light.

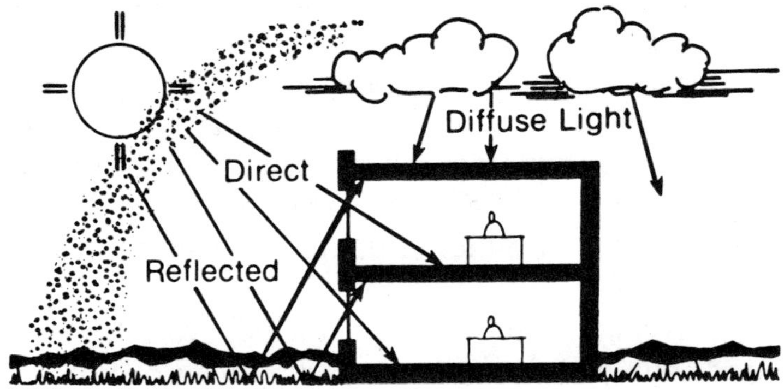

Fig. 7 — Sources of solar radiation.

Compared to the sun, the sky has a large visual area and a relatively low luminance. The amount of daylight received from the sky depends on the position of the sun and on atmospheric conditions. For design evaluation, one or more of three conditions of incident light on the exterior vertical surface of the building are usually considered. The first is light from an overcast sky only; the second is light from a clear sky only; and the third is light from a clear sky plus direct sunlight.

The amount of light received from an overcast sky depends on the luminance pattern of the sky

According to the Illuminating Engineering Society's handbook on daylight design, the amount of light received from an overcast sky and the direction from which this light reaches the windows of a building depend on the luminance pattern of the sky. The luminance distribution of an overcast sky varies with the location, time, density, and uniformity of the overcast. A uniformly overcast sky is normally 2.5 to 3 times as bright overhead as near the horizon. As a simplifying measure, a single luminance value representing the equivalent uniform sky luminance may be employed for design purposes.

The clear sky is normally brighter near the horizon than it is overhead

The sky luminance on clear days varies with the position of the sun and the amount of atmospheric dust or haze. Except in the immediate vicinity of the sun, the clear sky is normally brighter near the horizon than overhead. The concept of equivalent sky luminance mentioned above may also be used for clear skies. In daylight calculations for clear days, sky light only is included on non-sun exposures, while light from both sun and sky is included in calculations for sun exposures.

The illumination received on a horizontal surface from the clear sky alone is correlated with solar altitude. In the case of vertical or tilted surfaces, this illumination also varies with the orientation of the particular surface.

Sunlight reflected from the ground commonly represents 10 to 15 percent of the total daylight reaching a window area. Snow, light sandy soils, or light vegetation may exceed these levels. For non-sun exposures, the light reflected from the ground may account for more than half of the total light reaching the windows.

The light reflected from the ground may account for more than half of the light reaching the windows

Daylight is a variable, which may cause difficulties for the designer but is a psychological boon to people relying on it. The variations are somewhat predictable, in accordance with changing weather conditions and the daily and seasonal rhythms of the sun. There is, however, a lack of information about daylight availability in the United States. It is difficult to obtain information on whether the sky for a particular area is predominantly clear or overcast or data about the fraction of the working hours in a year during which one can expect minimum sky conditions to be exceeded.

Measurements that were made some time ago in Washington, D.C., and other urban areas are now suspect due to atmospheric changes resulting from air pollution and other climatic variables. Rather than wait several years before such data can be collected, it may be feasible to generate daylighting availability data from the solar radiation data now being collected throughout the United States. Efforts to test the validity of such techniques are now in progress.

Daylight design requires sensitivity beyond that provided by simple rules of thumb. The thermal/illumination tradeoffs and the qualitative as well as quantitative issues cannot be grossly simplified. A starting point, nonetheless, is to note that for pure daylighting purposes, the bigger the window opening, the better. The higher in the wall and the higher the ceiling, the better the design. If glass must be limited for thermal or comfort reasons, take it off the bottom for visual task performance and off the top for ambient purposes, where ground reflections can be useful. Openings are better horizontal than vertical because vertical openings set up strong patterns of light and dark contrasts. A predominant north/south orientation is useful because it lends itself to seasonal, or fixed, controls. Separation of view glass from daylight sources may help in making thermal and illumination design tradeoffs.

Daylight design requires sensitivity beyond that provided by simple rules of thumb

Sidelighting

The most familiar approach to daylight design is that of sidelighting

The most familiar approach to daylight design is sidelighting. Typically, this uses diffuse radiation only. Direct solar gain, although occasionally pleasant, often leads to visual and thermal discomfort, although it may be integrated into a passive solar heating design approach.

Daylighting levels from windows in one wall of a room fall off rapidly as we move deeper into the room. A typical practical limit for daylight penetration into an office is 15 to 20 feet from the window wall. Design approaches based on sidelighting approaches were well developed in the 1940's and 1950's; and the basic strategies for unilateral, bilateral, and trilateral daylighting designs have remained unchanged.

Fig. 8 illustrates classic unilateral daylighting schemes. It shows sections that were tested under the artificial sky at Rice University during the mid-1960's. The rooms were typical unilateral designs, 30 x 30 feet, and with 8-foot-high ceilings. Windows went from a sill height of three feet up to the ceiling; reflectances were 85 percent for the ceiling; 60 percent for the walls; 40 percent for the floor; 40 percent for the terrain. Test stations were located 30 inches above the floor, or at about desk height.

Unilateral lighting must be balanced or reinforced by additional daylight

Lighting intensities decrease from a high at the center of the window space to approximately one-fourth intensity at the center of the room and one-sixth intensity at the center of the back wall. The researchers concluded that unilateral lighting must be balanced or reinforced by additional daylight or artificial light to achieve reasonable uniformity for tasks located away from the windows.

The Rice researchers also tested a unilateral section with a succession of overhangs, from 4 feet to 8 feet. From a lighting standpoint, the rooms begin at the edge of the overhang; the location of the window wall merely defines the usable portion of the space below the roof. The lighting at the windows decreased substantially as the overhangs increased; lighting at the center and rear of the space was less affected. The uniformity of distribution was improved.

Illumination Levels — Unilateral Lighting

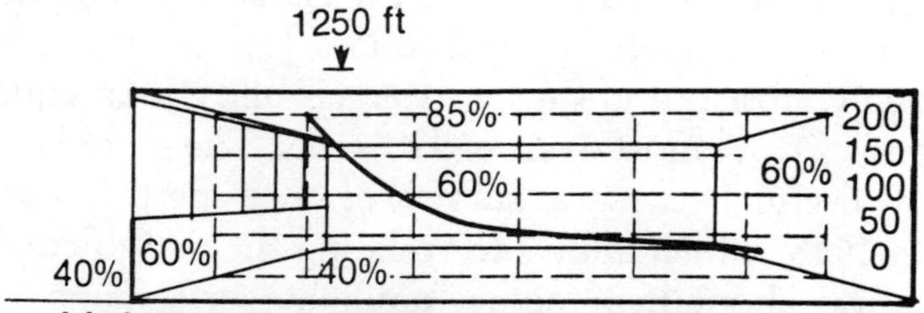

Lighting Gradient on the Center Line of Typical Room

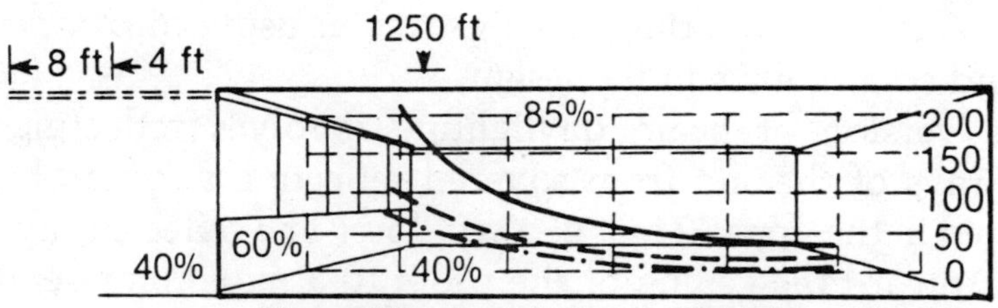

The Effects of Roof Overhangs on Lighting Distribution

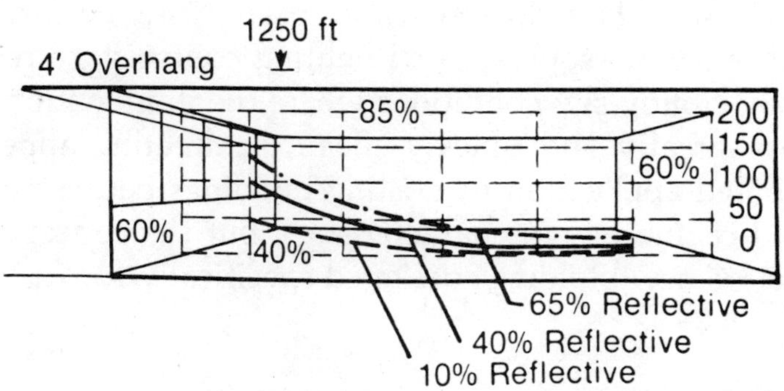

Gradients Showing Changes Due to Terrain Reflectivities

Fig. 8 — Tests under artificial sky by Rice University.

Effective sidelighting produces greater visibility than equivalent footcandles from overhead systems

The California schools designed in the 1940's and 1950's to use daylight are a good reference for classic bilateral daylight design. Effective sidelighting produces less ceiling reflection, improved contrast, and thus greater visibility than equivalent foot-candles from most overhead systems. The schools are particularly useful because a survey of their performance was made in 1948, which allows for relating the design concept to information about their actual daylight performance. **Fig. 9** shows generic diagrams of classic multilateral daylight building sections, examples of those sections as used in the schools, and reading of light levels on their interiors.

Some techniques for increasing daylight penetration, which either allow a wider building section or a greater percentage of daylight reliance are being closely examined today. The most discussed strategy is beam daylighting, of which the Tennessee Valley Authority building is an example, using mirror reflectors and solar courts in its design.

The control of reflected light as sun angles change can be a problem

One version of beam daylighting involves reflecting the direct rays of the sun from silvered reflective venetian blinds mounted in the upper 2 feet of a window. The reflected rays are aimed toward the ceiling of the room to a maximum depth of 30 to 40 feet. The ceiling then acts as a diffuse reflector providing normal diffuse illumination deep inside the room. Although the light quality achieved by such a scheme is satisfactory, the control of reflected light as sun angles change can be a problem.

Beam sunlighting appears to have limited application to existing buildings

A variety of controllable reflecting or refractor-type devices has been examined, but the real issue is achieving simplicity and low cost in these devices without sacrificing performance. In addition, more sophisticated lighting controls are required for partly cloudy sky conditions where the illumination levels will vary greatly and quickly. Beam sunlighting appears to have limited application to existing buildings because of window and ceiling design characteristics, but the concept offers promise for new buildings designed specifically for it.

Toplighting

Toplighting is another important daylight design strategy, promising more uniform light for general illumination, although less appropriate light for tasks. Historically, skylights have been designed to exclude the direct sun (sawtooth and skylights with overhangs are examples).

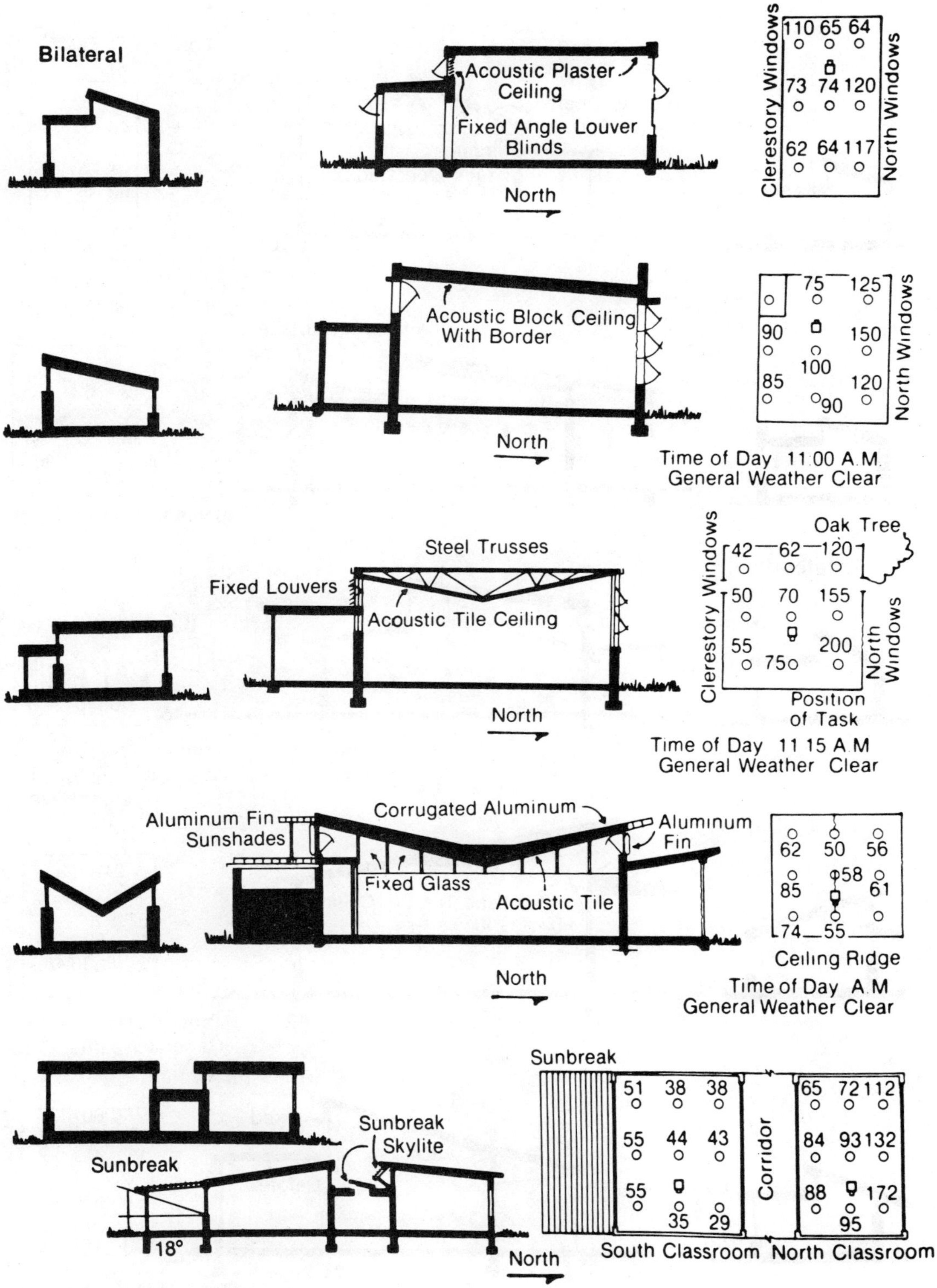

Fig. 9 — Lighting levels for various building types.

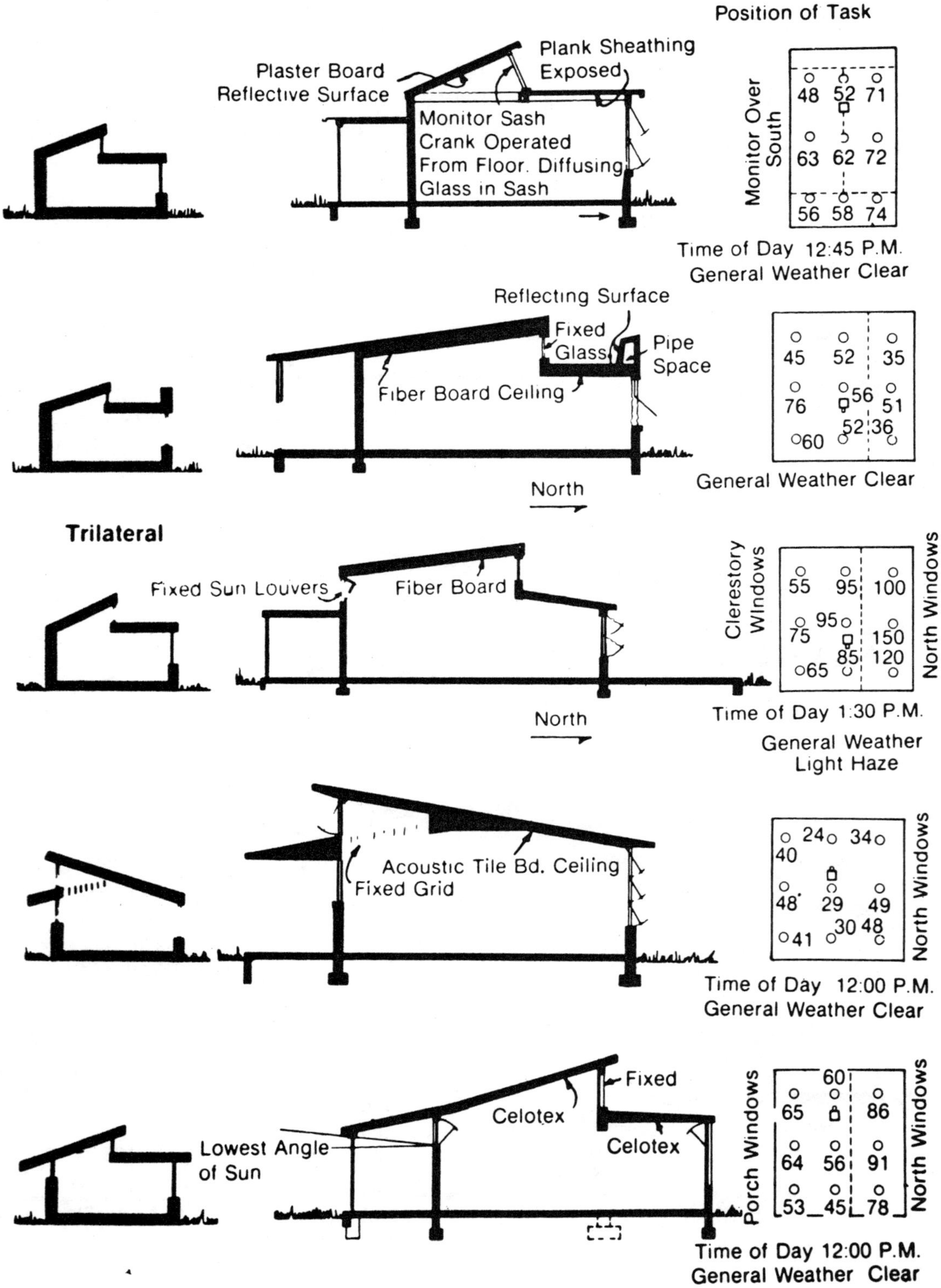

Fig. 9 — Lighting levels for various building types.

Large spaces, such as factories and warehouses, lend themselves to variants of conventional sidelighting and toplighting designs, such as roof monitors and sawtooth clerestories. More recently, translucent roofs provide substantial access to daylight, although control becomes difficult. An analysis of some industrial applications was published by Detroit Steel Products Co. in 1929.

Translucent roofs provide substantial access to daylight, although control becomes difficult

Fig. 10 compares the daylight performances of vertical versus sloped clerestories at the Bethlehem Steel Co. An increase in monitor height increased performance, although the researchers concluded that 8-foot-high sloping windows admitted no more light than vertical windows 6 feet, 10 inches high, as dirt accumulated over a six-month period.

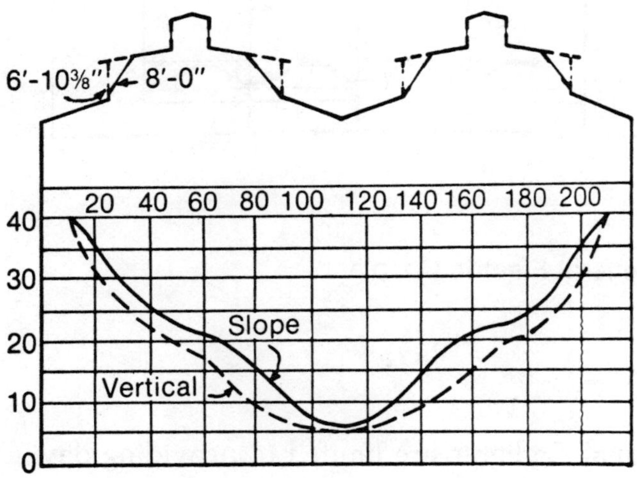

Fig. 10 — Bethlehem Steel Co.

Domed or peaked skylights provide a larger window to the sky and thus more daylight

The most successful example of natural lighting in this study was a General Electric plant, 340 feet long, with a sawtooth span of 25 feet. This design gained high illumination and uniformity (Fig. 11).

Most modern skylights lie flat in the roof plane and thus accept both direct and diffuse radiation. Domed or peaked skylights provide a larger window to the sky and thus more daylight. All skylights raise the issue of heat along with light, and controls by orientation or insulation may be needed.

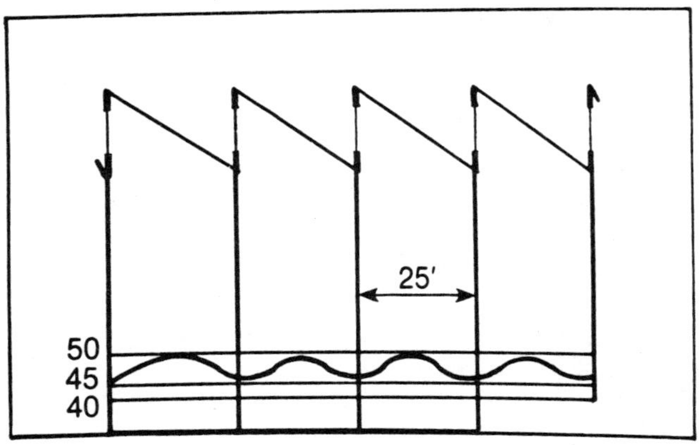

Fig. 11 — General Electric Co. plant.

Conventional skylights are limited to providing daylight in one-story or multi-atrium spaces, but several concepts recently have been explored that use direct sunlight through a roof aperture and then distribute that light through the building by using mirrors, lenses, and other optical controls. Such schemes are feasible, but their optical performance requirements and complexity limit their practicality for most building applications.

Orientation is a significant factor in skylight design. Several recent strategies have sought dynamic adaptation to heating and cooling conditions by developing movable insulation systems within a skylight. A north-facing skylight accepts light but does not produce much solar heat gain; a south-facing skylight admits both heat and light. Movable insulation can allow the designer to take advantage of both at appropriate times.

Design analysis

A major obstacle to daylighting design is a continuing lack of simple and accurate design analysis methods. These methods can be computational or graphic solutions; they can be in the form of graphic tables and computer programs; they include protractors and physical models. All of these exist, but none provides the combination of reliability, accuracy, and dynamic response that daylight design requires. The designer must therefore choose among them according to specific program and design objectives and use the results with caution.

All calculations are at best a representation, or simplification, of reality. Computer programs are a model of reality; what they do and do not measure is frequently more important than how well they measure a given element. The daylight designer must therefore approach all such programs and calculations as useful but not conclusive information. Daylight is a visual phenomenon; one should believe one's eyes and use them to challenge calculation. Because daylight is an instantaneous event and cannot be stored, averages can be very misleading.

All calculations are at best a representation, or simplification, of reality

Physical modeling of a daylight design is therefore critical. Such a model is valuable for predicting both the quality and quantity of daylight and can reveal lighting gradients, specific glare problems, and the effects of building forms and finishes. Whereas a mathematical model may work for a standard design condition, a physical model will provide information on complex, unusual, and specific design performance.

A physical model will provide information on complex, unusual, and specific design performance

After being built, the model should be tested outdoors, under naturally varying conditions. An artificial sky allows testing under more controlled conditions but implies the kind of stability and exact numbers that belie actual daylight conditions. Preferably, the model should be tested at the proposed building site so that it can demonstrate such local conditions as reflection from landscaping or adjacent buildings. Outdoor testing suggests the limitations of real time, but the passage of seasons can be simulated by tilting the model to obtain the proper relationship to the sun.

A daylight model can be "quick and dirty," compared to a conventional architectural model. In the early stages of design, the model may simply deal with basic dimensions, openings, and surfaces; later in the design process, specific architectural features and furnishings must be modeled.

Some model materials will transmit light and must be covered to make them completely opaque

The model must be large enough to use with a meter and camera or other instruments, but small enough to carry to the site (approximately 1:20 scale). A model need not be painstakingly detailed; it needs to be accurate only in terms of room proportions, opening sizes, and reflectance and transmission values of plus or minus 10 percent. Be aware that some model materials will transmit light and must be covered with chipboard, aluminum foil, or paint to make them completely opaque. If adjacent surfaces, such as roofs and pavement, reflect enough light into the model to affect measurements, they should be included in the model and have an accurate reflectance. The plan orientation of the space in relation to north should be noted, and the model should have hatches in the floor and walls for taking measurements and photographs. If there is no glazing in the model itself, measurements must be factored to reflect actual glazing conditions.

All but a miniscule fraction of buildings are daylit buildings

Numerical calculations complement the physical model design evaluation. They allow for comparison and tradeoffs and an overall assessment of design performance. All but a miniscule fraction of buildings are daylit buildings. But not all of them are good daylit buildings that use daylight to provide useful illumination, an environment of good quality, and energy efficiency. Ask most architects to design a room to provide 50 footcandles on a desk for 80% of working hours by using daylight and you are likely to get a puzzled look and a fast shuffle through papers and textbooks. The problem is not a lack of methods but a lack of accessible, simple, and appropriate methods.

Design methods

There are, in fact, about 60 recognized daylight design methods. But most were developed in Europe, where overcast skies prevail, and few are well suited to this country, where clear skies prevail. All of these techniques, however, can be divided into two main categories: the lumen method and the daylight factor method.

The lumen method was developed by J.W. Griffith at Southern Methodist University for Libbey-Owens-Ford Co. on the basis of extensive model studies under an artificial sky. The method was adopted by the Illuminating Engineering Society and has been the most widely used method in the United States.

The lumen method is applicable to a limited range of window configurations

The lumen method allows a comparison of various window wall schemes and their economic tradeoffs with HVAC and electrical lighting values. The effects of various glass transmittances, ground reflectances, wall reflectances, diffusing shades, venetian blinds, overhangs, clear and overcast skies, and direct solar radiation are included in the calculation, which can evaluate them for energy efficiency and return on investment.

The method is limited to schemes of maximum utilization of daylight, and it predicts illumination levels on a center line from the windows at intervals of five feet from the back of the room. The technique was designed to evaluate schemes of variable window management, as well as fixed controls, to obtain the total benefits of daylight utilization, rather than meeting a minimum requirement. The method is applicable to a limited range of window configurations and accurate only for points situated along the center line from the window; it is also limited to vertical glazing.

The daylight factor

The daylight factor method was developed in Great Britain to meet a minimum code requirement. It predicts the daylight anywhere in the room for a standard sky condition without sun incident on the window. The daylight factor is described as a percentage of diffuse light received at a point of reference. The percentage expresses the ratio of interior illumination at a given point to the available illumination from an unobstructed sky outdoors. The point of reference is a horizontal plane.

The daylight factor is described as a percentage of diffuse light received at a point of reference

This method allows nonuniform sky luminance distribution and is sensitive to the position of a window in a wall. There is no limit to room size or shape, and it accommodates nonvertical glazing. Its capacity to respond to shading devices is limited, as is its capability to deal with clear sky conditions. Direct sun is essentially excluded. The daylight factor, however, does provide a relative measure of all daylight, not only light directly from the sky (including losses from glass, dirt, window bars, etc.), but also the contributions from light reflected directly to a reference point from visible exterior sur-

The daylight factor does provide a relative measure of all daylight

faces and from light interreflected between the room's interior surfaces.

Both the lumen and daylight factor methods can be used in the early stages of design to evaluate alternative schemes without a computer. If the window schemes are on the east, west, or south, the lumen method allows tradeoff evaluations with variable sun controls.

Computer models have also been developed for daylight calculations

Computer models have also been developed for daylight calculations: they compute illumination levels and equivalent sphere illumination (ESI) levels in daylit rooms. (ESI is a useful or nonglare foot-candle. Specifically, it is the foot-candle illumination in a sphere that gives equal visual performance to the illumination, of whatever foot-candles, in a specific lighting system. Daylight from side windows provides two or three times more ESI than typical overhead light sources. ESI cannot be effectively measured in a physical daylight model because the measurement instruments are too large.) Available computer models, however, cannot easily model specific architectural details or such information as the placement and nature of room furnishings. The computer models also must still rely on essentially static analysis of a dynamic design problem.

Lumen II Daylight is a computer program for predicting daylight illumination and ESI for a simple task anywhere in the room. It is relatively expensive and therefore of limited usefulness in the early stages of design.

Daylighting and energy performance

The physical and mathematical daylight analysis methods discussed deal only with the illumination aspects of daylight design. Daylight implies thermal tradeoffs that also need evaluation. The sizing and placement of windows to maximize daylighting benefits need to be considered in the context of thermal performance and overall energy performance.

Daylight implies thermal tradeoffs that also need evaluation

A number of computer programs have been developed for energy analysis, but they are based on certain mechanical assumptions about the thermal design of a building. And none deals with daylight. Most concentrate on thermal design and deal with illumination by using standardized operating profiles for installed lighting.

Some of these programs are relatively complex and can model a building's performance hour by hour throughout the year. Daylight performance has been incorporated experimentally into several of these, but the results have yet to be validated by comparison with actual conditions. None is at a state where an architect could use it with confidence to obtain accurate information.

The pressures for an integrated thermal/illumination model incorporating daylight continue to build. New energy performance standards will generate more pressure to incorporate daylight into existing energy analysis programs. But the problems of validation and the implications of the simplified data used in a computer program have yet to be resolved.

The issues surrounding the illumination aspects of daylight and how to provide the designer with accurate information continue also to be the subject of intensive research. Daylight availability is so variable that trying to predict absolute values is not cost effective; therefore the precision of any measurement may be more or less important depending on the particular case. The lumen or daylight factor methods will give the architect a relative evaluation of alternative schemes useful for early design decisions. The physical model remains the most cost-effective way to approach absolute values for specific design solutions, allowing both qualitative and quantitative evaluation and perceptual as well as analytical insights.

The physical model remains the most cost-effective way to approach absolute values for design solutions

Quicklite 1

Quicklite 1 will predict daylight illumination at any point within a room. The program uses the CIE sky luminance distribution functions for the overcast and clear skies. The light reaching the interior point being considered is separated into two components. Light arriving directly from the sky (Sky Component or SC) is calculated using a source area formula. Light reflected from external and internal surfaces (Reflected Component or RC) is calculated using the split flux approach. The total for these two components is given as either the daylight factor (the ratio of illumination at the reference point to illumination on a horizontal surface from the unobstructed sky) or illumination (in footcandles) for the point(s) being considered. The following illustrates the use of the program to calculate daylight illumination for a simple window, without obstructions or sun-control devices. The program can also be used to solve more complex daylighting problems.

Quicklite 1 will predict daylight illumination at any point within a room

Quicklite 1 has been designed to be accurate, simple, and relatively fast-running. The output compares well with other daylighting calculation procedures as well as with a series of physical scale-model measurements. However, it must be remembered that no calculation procedure, no matter how accurate, can duplicate all the complexities that daylighting poses. In most cases we seek comparative performance levels rather than absolute quantitative results. Remember that the non-quantifiable aspects of lighting design are equally important and that nothing can substitute for good design judgment.

Procedures

Sky component program: This program uses the TI-59's master library. If multiple point calculations are desired for a single program execution, the PC-100 printer must be used.

Step 1: Repartition the calculator memory by entering 5 and then pressing 2nd OP 17.

Step 2: Determine if sky condition is clear or overcast, then read appropriate program into calculator. The clear sky program will require reading sides 1, 2, and 3, while the overcast sky program requires reading only sides 1 and 2.

Step 3: a) Enter window transmission at normal incidence (90°to the plane of the window).

NOTE: Steps should be executed only for the clear sky program.

b) Enter the normalization factor (N_{sc}), which is the ratio of illumination on a horizontal plane to the luminance at the zenith. This factor is a function of solar altitude and is given as follows:

Solar Altitude	N_{sc}
10°	3.654
20°	3.565
30°	3.235
40°	2.741
50°	2.178
60°	1.632
70°	1.159
80°	0.782
90°	0.503

c) Enter window azimuth angle (in degrees), which is the angle that a line normal to a vertical window makes with south (which is taken to be 0°). Clockwise rotation from south is measured as positive while counter-clockwise rotation is measured as negative (**Figure 12**).

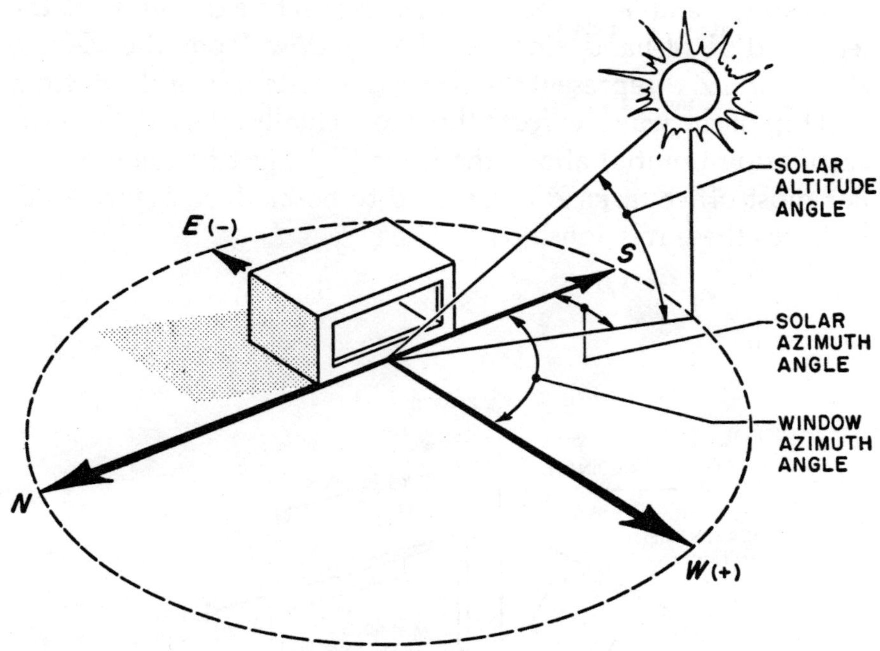

Fig. 12 — Site coordinate system for program.

d) Enter solar azimuth angle (in degrees), which is the azimuthal angle of the sun from south. Here again clockwise rotation from south is measured as positive while counter-clockwise rotation is measured as negative (Figure 12).
NOTE: This program does not compute the effect of direct sun in the room. There will be direct sun at some point in the room for an unobstructed window if the absolute value of the difference between the window azimuth and solar azimuth is less than 90°.
e) Enter solar altitude [in degrees from horizontal], (Figure 12).

Step 4: Enter the location and size of the window and the workplane height at which illumination is to be calculated. A rectangular coordinate system is used to locate the window as well as the calculation point(s) (**Figure 13**). The origin of this coordinate system should be placed at the lower left-hand corner of the window-wall's wall. The Z-axis will point straight up. The location and size of the window is described by four values: Y_{left} and Y_{right} represent the respective distances of the left- and right-hand sides of the window from the Z-axis, Z_{bottom} and Z_{top} represent the respective distances of the bottom and top of the window from the floor. Finally, the height of the calculation point(s) above the floor (Z_p) must be determined. For most office tasks Z_p is assumed to be 2.5 feet. Figure 13 illustrates these relationships.

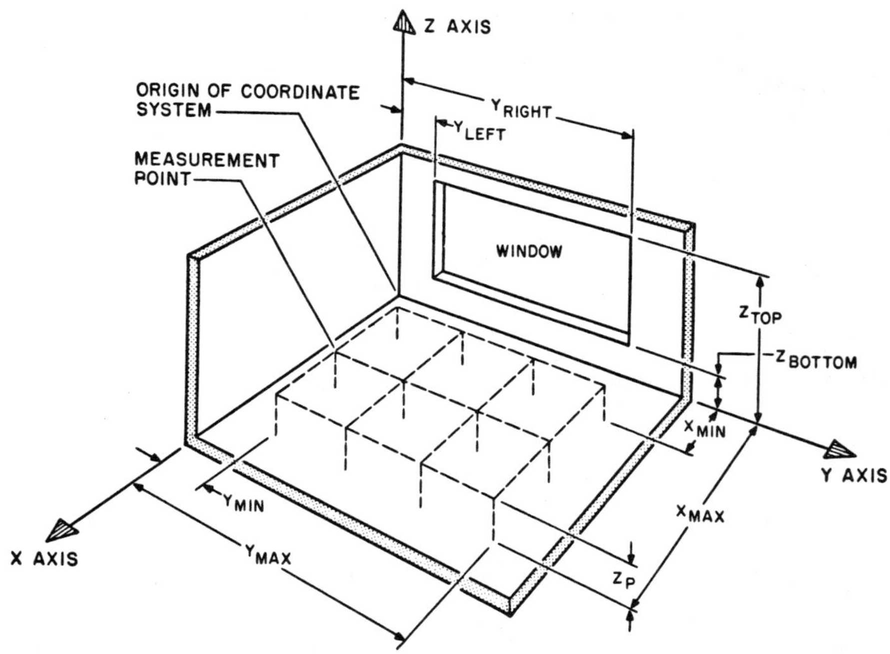

Fig. 13 — Room coordinate system for program.

Step 5: Determine if the output is to be calculated in units of daylight factor or illumination. To calculate the daylight factor, the number 1 must be stored in register 35. To calculate illumination, the appropriate horizontal illumination from the sky (E_{sky}) must be stored in register 35. This value can be found in **Figures 14/15**, or can be taken from any other source of measured or calculated illumination data.

NOTE: To calculate values for a single point, proceed to steps *6a* and *7a*. If a printer is available, multiple points on a rectangular grid may be calculated and printed by proceeding to steps *6b* and *7b*.

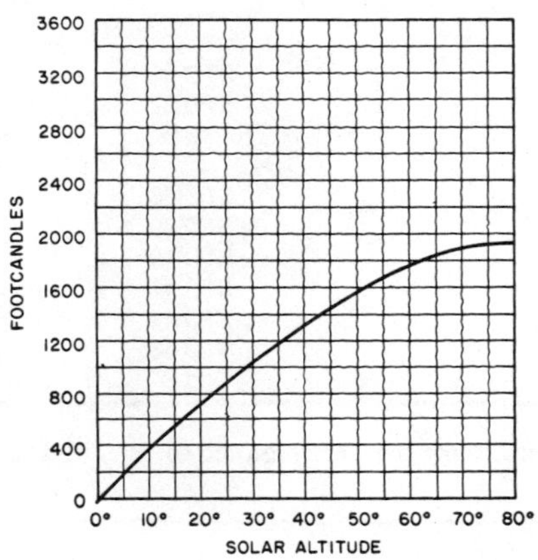

Fig. 14 — Overcast sky illumination on a horizontal surface (E_{sky}). Curve assumes a temperate (humidity = 0.3) urban (turbidity = 0.1) environment.

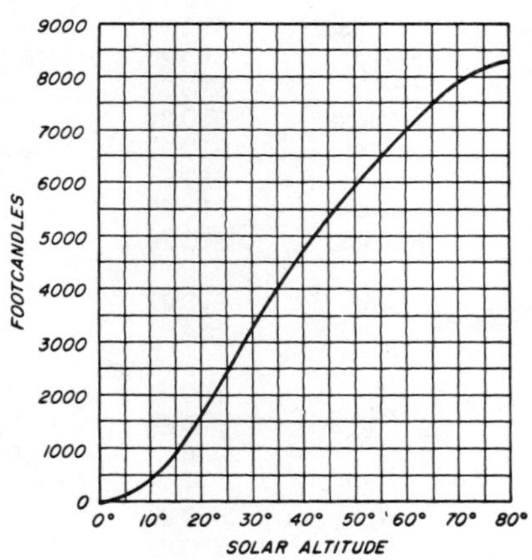

Fig. 15 — Clear sky illumination on a horizontal surface (E_{sky}). Curve assumes a temperate (humidity = 0.3) urban (turbidity = 0.1) environment.

Sky component program worksheet, summarizing the steps required to enter and run the sky component program.

Sky component program worksheet

Step	Procedure	Enter	Press	Display
1	Repartition calculator	5	2nd OP 17	559.49
2	Clear sky: read sides 1, 2, & 3	1,2,3	2nd INV WRITE	1,2,3
	Overcast sky: read sides 1 & 2	1,2	2nd INV WRITE	1,2
3a	Enter window transmission	T_0	A	$1.018T_0$
Note:	Steps 3b–3e should be executed for clear sky only			
3b	Enter normalization factor	N_{sc}	R/S	N_{sc}
3c	Enter window azimuth (in degrees)	AZ_{window}	R/S	AZ_{window}
3d	Enter solar azimuth (in degrees)	AZ_{solar}	R/S	AZ_{solar}
3e	Enter solar altitude (in degrees)	ALT_{solar}	R/S	ALT_{solar}
4	Enter Y_{left}	Y_{left}	STO 30	Y_{left}
	Enter Y_{right}	Y_{right}	STO 31	Y_{right}
	Enter Z_{bottom}	Z_{bottom}	STO 32	Z_{bottom}
	Enter Z_{top}	Z_{top}	STO 33	Z_{top}
	Enter Z_p	Z_p	STO 34	Z_p
5	To calculate daylight factor: enter 1 To calculate illumination: enter E_{sky}	1 or E_{sky}	STO 35	1 or E_{sky}
Note:	For single point calculation:			
6a	Enter Y_p	Y_p	STO 36	Y_p
	Enter X_p	X_p	STO 37	X_p
7a	Run program	—	2nd B′	S.C.
Note:	For multiple point calculations:			
6b	Enter Y_{min}	Y_{min}	STO 38	Y_{min}
	Enter Y_{max}	Y_{max}	STO 39	Y_{max}
	Enter X_{min}	X_{min}	STO 40	X_{min}
	Enter X_{max}	X_{max}	STO 41	X_{max}
	Enter number of points along Y axis	N_y	STO 42	N_y
	Enter number of points along X axis	N_x	STO 43	N_x
7b	Run program	—	2nd D′	S.C.

Reflected component program worksheet, summarizing the steps required to enter and run the reflected component program.

Reflected component program

Step	Procedure	Enter	Press	Display
1	Repartition calculator	3	2nd OP 17	719.29
2	Read sides 1 & 2	1,2	2nd INV WRITE	1,2
3	Enter window width	$Width_{win}$	STO 15	$Width_{win}$
	Enter window height	$Height_{win}$	STO 16	$Height_{win}$
4	Enter room width	$Width_{rm}$	STO 17	$Width_{rm}$
	Enter room depth	$Depth_{rm}$	STO 18	$Depth_{rm}$
	Enter room height	$Height_{rm}$	STO 19	$Height_{rm}$
	Enter window sill height	$Height_{sill}$	STO 20	$Height_{sill}$
5	Enter window transmission	T_0	STO 21	T_0
6	Enter ceiling reflectance	ρ ceiling	STO 22	ρ ceiling
	Enter floor reflectance	ρ floor	STO 23	ρ floor
	Enter glass reflectance	ρ glass	STO 24	ρ glass
	Enter wall reflectance	ρ wall	STO 25	ρ wall
	Enter ground reflectance	ρ ground	STO 26	ρ ground
7a	Daylight factor for an overcast sky:			
	Enter window factor	f_s	STO 27	f_s
	Run program	—	A	R.C.
7b	Illumination for an overcast sky:			
	Enter window factor	f_s	STO 27	f_s
	Enter illumination from sky	E_{sky}	STO 28	E_{sky}
	Run program	—	B	R.C.
7c	Daylight factor for a clear sky:			
	Enter window factor	f_s	STO 27	f_s
	Enter illumination from sky	E_{sky}	STO 28	E_{sky}
	Enter illumination from sun	E_{sun}	STO 29	E_{sun}
	Run program	—	C	R.C.
7d	Illumination for a clear sky:			
	Enter window factor	f_s	STO 27	f_s
	Enter illumination from sky	E_{sky}	STO 28	E_{sky}
	Enter illumination from sun	E_{sun}	STO 29	E_{sun}
	Run program	—	D	R.C.

Single point procedure:

Step 6a: Enter the location of the point at which the daylight factor of illumination calculation is to be made. Y_p is the distance from the measurement point to the wall to the left of the window. X_p is the distance from the measurement point to the window-wall (Figure 13).

Step 7a: Calculate the sky component by pressing 2nd B′.

Multiple point procedure:

Step 6b: Enter the corner locations of the rectangular grid of calculation points. Y_{min} and Y_{max} are, respectively, the distances from the wall at the left of the window to the closest and the farthest columns of calculation points. X_{min} and X_{max} are, respectively, the distances from the window-wall to the closest and the farthest rows of calculation points (Figure 13). Finally, the number of calculation points for the Y-column (N_y) and X-row (N_x), must be determined. N_y and N_x include the corner points and thus must be greater than or equal to two. Setting N_y or N_x to one will result in a "divide by zero" error.

Step 7b: Calculate the sky component values by pressing 2nd D′. Run time for the option is approximately 1.2 minutes per point for overcast sky and 3 minutes per point for clear sky.

General notes: sky component program

1) It is unnecessary to re-enter all the input data for repeated calculations. Only those input values entered in step *6a* or *6b* must be re-entered to rerun the program. All other input values need to be changed only if their new values differ from those used in the previous run.

2) All inputs that relate to distance (such as width of the window) may be entered in any unit, English or metric, so long as the user is consistent.

3) If a printer is being used, it is highly advisable to make a list of the register contents before running the program. To obtain a permanent list of the data registers, enter 30 and then press INV LST before executing step *7a* or *7b*. (For register assignments see **Figure 18**).

Reflected component program: This program uses the TI-59's master library. The PC-100 printer is optional.

Step 1: Repartition the calculator memory by entering 3 and then pressing 2nd OP 17.

Step 2: The reflected component requires reading sides 1 and 2.

Step 3: Enter window width and height.

Step 4: Enter room width, depth, and height, plus the height of the window sill above the floor.

Step 5: Enter normal incidence window transmission.

Step 6: Enter reflectance of ceiling, floor, glass, and walls, as well as exterior ground.

NOTE: There are four options for output from this program. The appropriate option must be chosen from steps *7a, b, c,* or *d*.

Step 7a: To calculate the daylight factor from the reflected component under an overcast sky, enter the window factor (f_s). The window factor is the ratio of the light flux entering the window from the sky to the light flux on an exterior horizontal plane. For an unobstructed sky $f_s = 0.39$.

Step 7b: To calculate the illumination level from the reflected component under an overcast sky, enter the window factor (f_s) as in step *7a* and the horizontal illumination from the sky (E_{sky}), which can be found using Figure 14.

Step 7c: To calculate the daylight factor from the reflected component under a clear sky, enter the window factor (f_s), the horizontal illumination from the sky (E_{sky}) from Figure 15, and the horizontal illumination from the sun (E_{sun}) from **Figure 16.** The window factor for a clear sky is a function of solar altitude as well as of window azimuth from the sun and is given below.

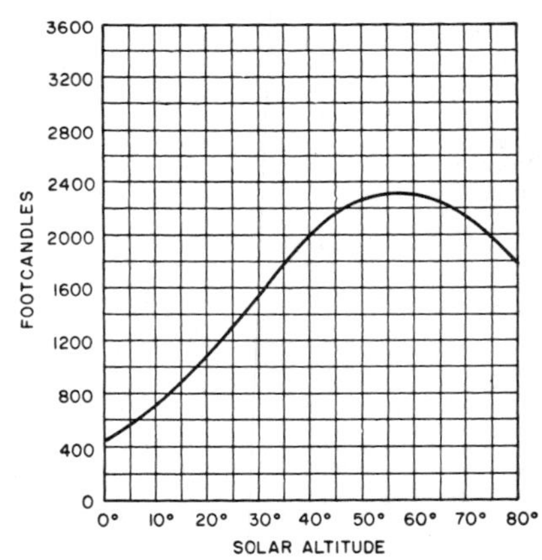

Fig. 16 — Direct solar illumination on a horizontal surface (E_{sun}). Curve assumes a temperate (humidity = 0.3) urban (turbidity = 0.1) environment.

Variable	Overcast Sky Input	Overcast Sky Memory #	Clear Sky Input	Clear Sky Memory #
Y_{left}	1.	30	1.	30
Y_{right}	29.	31	29.	31
Z_{bottom}	2.	32	2.	32
Z_{top}	7.	33	7.	33
Z_p	2.5	34	2.5	34
E_{sky}	1325.	35	2000.	35
Y_p	0.	36	0.	36
X_p	0.	37	0.	37
Y_{min}	5.	38	5.	38
Y_{max}	25.	39	25.	39
X_{min}	5.	40	5.	40
X_{max}	15.	41	15.	41
N_y	3.	42	3.	42
N_x	3.	43	3.	43
$1.018T_0$	0.79404	44	0.79404	44
N_{sc}	0.	45	2.741	45
AZ_{window}	0.	46	90.	46
AZ_{solar}	0.	47	0.	47
ALT_{solar}	0.	48	40.	48
	0.	49	0.	49

Output

Overcast Sky		Clear Sky	
5.	X	5.	X
5.	Y	5.	Y
84.19540063	SC	129.81096	SC
5.	X	5.	X
15.	Y	15.	Y
97.26722096	SC	174.7760171	SC
5.	X	5.	X
25.	Y	25.	Y
84.19540063	SC	161.3710915	SC
10.	X	10.	X
5.	Y	5.	Y
20.37312296	SC	43.07689806	SC
10.	X	10.	X
15.	Y	15.	Y
25.82729313	SC	62.55798576	SC
10.	X	10.	X
25.	Y	25.	Y
20.37312296	SC	54.75495501	SC
15.	X	15.	X
5.	Y	5.	Y
7.796449167	SC	20.07071398	SC
15.	X	15.	X
15.	Y	15.	Y
9.922776182	SC	28.0289317	SC
15.	X	15.	X
25.	Y	25.	Y
7.796449167	SC	25.03791439	SC

Fig. 18 — Input and output for sky component program.

Step 7d: To calculate the illumination level from the reflected component under a clear sky, proceed as in step *7c*, but a different label is pressed for the resulting answer.

General notes: reflected component

1) A single average value for the reflected component is computed for the entire room.

2) It is unnecessary to re-enter all the input data for repeated calculations. Change only those values entered in steps *3* through *6* that differ from the previous run. Then execute the program beginning at step *7*.

3) If a printer is being used, it is highly advisable to make a list of the register contents before running the program. To obtain a permanent record of the data registers, enter 15 and then press INV LST before executing step *7*. (For register assignments see **Figure 19.**)

Variable	Overcast Sky Input	Overcast Sky Memory #	Clear Sky Input	Clear Sky Memory #
$Width_{win}$	28.	15	28.	15
$Height_{win}$	5.	16	5.	16
$Width_{rm}$	30.	17	30.	17
$Depth_{rm}$	20.	18	20.	18
$Height_{rm}$	8.67	19	8.67	19
$Height_{sill}$	2.	20	2.	20
T_0	0.78	21	0.78	21
$\rho_{ceiling}$	0.8	22	0.8	22
ρ_{floor}	0.2	23	0.2	23
ρ_{glass}	0.14	24	0.14	24
ρ_{wall}	0.5	25	0.5	25
ρ_{ground}	0.2	26	0.2	26
FS	0.39	27	0.5	27
E_{sky}	1325.	28	2000.	28
E_{sun}	0.	29	4750.	29
Output R.C.	25.	RC	78.	RC

Fig. 19 — Input and output for reflected component program.

Sample problem

Calculate the illumination for the room shown in **Figure 17**, which is 30 feet wide, 20 feet deep, and 8.67 feet high, and which has one large window, 5 feet high by 28 feet wide. The window is 2 feet above the floor and centered on a wall facing due west. Glazing consists of two layers of ¼-inch clear glass having a combined normal daylight transmission of 78 percent. The ceiling has a reflectance of 80 percent, the floor 20 percent, the wall 50 percent, and the glass 14 percent.

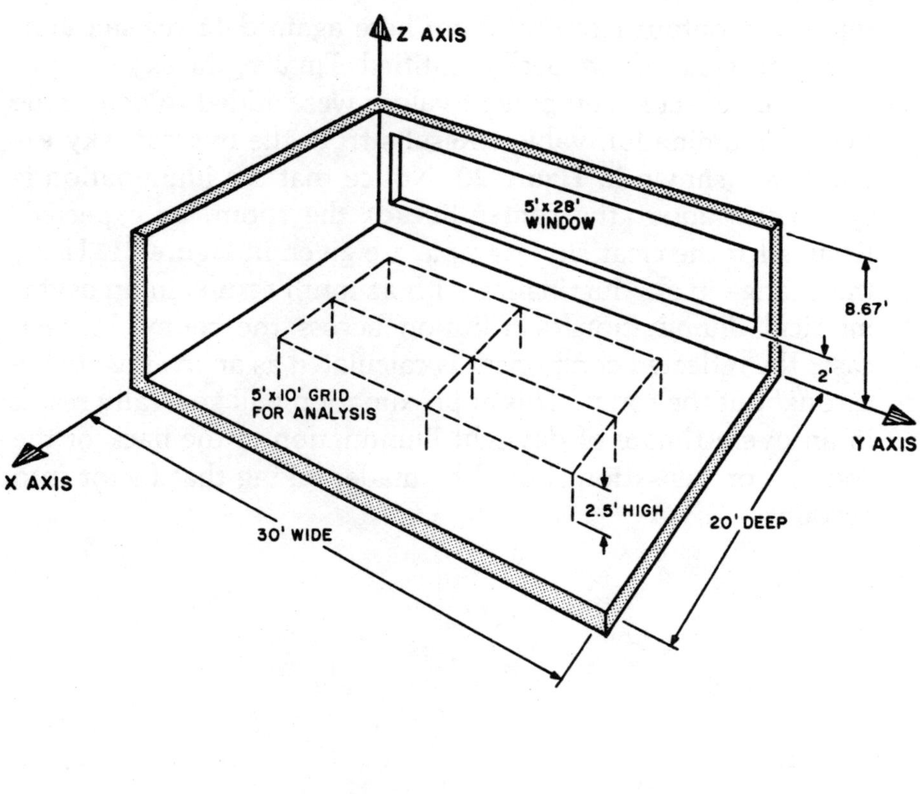

Fig. 17 — Example room.

Illumination is to be calculated under both clear and overcast conditions. The sun is assumed to be due south at an altitude of 40°. The horizontal illumination from the sky (E_{sky}) is 1325 footcandles under overcast sky conditions (from Figure 14) and 2000 footcandles under clear sky conditions (from Figure 15). The horizontal illumination from the sun (E_{sun}) is 4750 footcandles (from Figure 16). Finally, a three-by-three calculation grid (9 calculation points) was selected, each edge of the grid being five feet from the nearest wall (Figure 17).

The sky component program for both clear and overcast skies was run first. Figure 18 shows the input and output from this example in which both the data register contents and results are identified. The reflected component program for both sky conditions were then run. Figure 19 shows the sample input and output from this run; here again data register contents and results have been identified. Finally, the sky component and reflected component values were added to determine overall illumination values. Results from the overcast sky example are shown in **Figure 20**. Notice that the illumination is symmetric about the center line of the room, as expected. Results for the clear sky example are given in **Figure 21**. Here, the change in sky luminance with azimuth results in an asymmetrical illumination distribution across the room. In each case, the reflected component is calculated as an average value throughout the room. This approximation will generally result in an overestimate of daylight illumination in the back of the room. Compensation should be made, taking that factor into account.

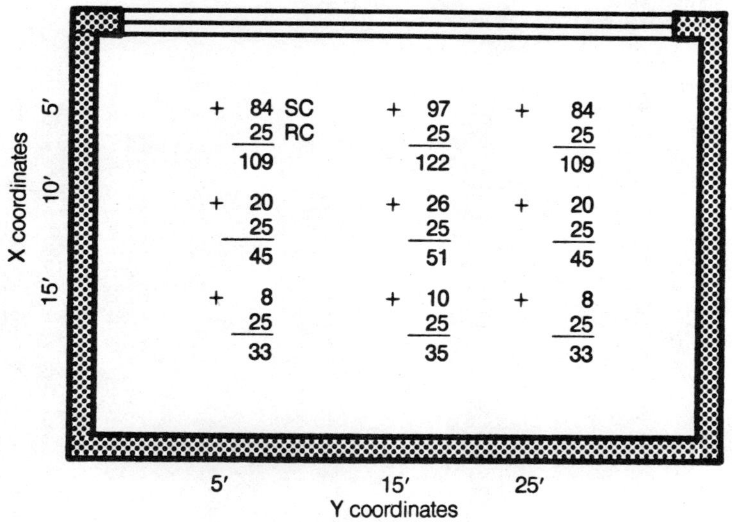

Fig. 20 — Overcast sky illumination values (in footcandles) for example room.

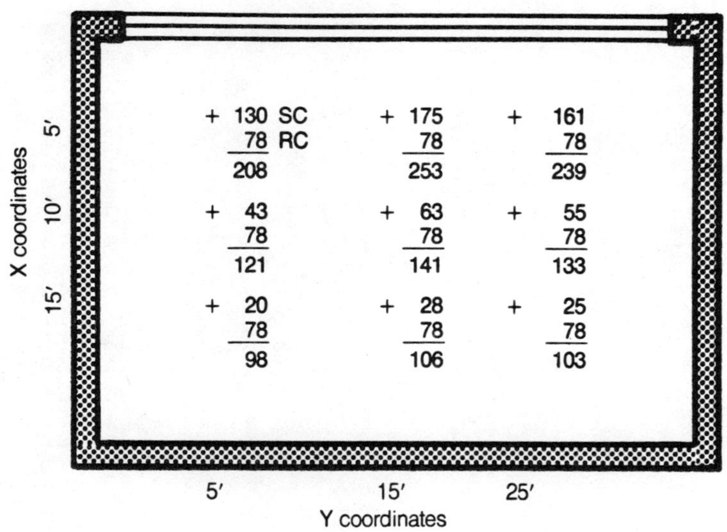

Fig. 21 — Clear sky illumination values (in footcandles) for example room.

Fig. 22 — Program listings.

Program Listings

Sky Component Program: Overcast Sky

```
76 LBL
16 A'
42 STO
15 15
39 COS
42 STO
16 16
33 X^2
53 (
42 STO
17 17
65 ×
43 RCL
12 12
85 +
43 RCL
13 13
54 )
42 STO
18 18
53 (
53 (
53 (
53 (
43 RCL
13 13
85 +
43 RCL
17 17
65 ×
43 RCL
14 14
54 )
55 ÷
43 RCL
18 18
54 )
45 Y^X
01 1
93 .
05 5
85 +
01 1
54 )
65 ×
53 (
02 2
65 ×
43 RCL
16 16
65 ×
43 RCL
11 11
55 ÷
43 RCL
18 18
34 √X
85 +
01 1
54 )
65 ×
43 RCL
17 17
65 ×
43 RCL
16 16
55 ÷
53 (
43 RCL
18 18
45 Y^X
02 2
93 .
05 5
54 )
54 )
92 RTN
76 LBL
11 A
65 ×
01 1
93 .
00 0
01 1
08 8
95 =
42 STO
44 44
91 R/S
76 LBL
13 C
43 RCL
23 23
42 STO
10 10
94 +/-
85 +
43 RCL
21 21
95 =
55 ÷
04 4
42 STO
09 09
95 =
42 STO
22 22
00 0
42 STO
19 19
42 STO
08 08
02 2
42 STO
07 07
76 LBL
58 FIX
03 3
42 STO
24 24
76 LBL
42 STO
02 2
22 INV
44 SUM
24 24
43 RCL
10 10
39 COS
42 STO
11 11
33 X^2
42 STO
12 12
42 STO
14 14
43 RCL
10 10
38 SIN
33 X^2
42 STO
13 13
94 +/-
44 SUM
14 14
04 4
36 PGM
09 09
13 C
36 PGM
09 09
14 D
65 ×
43 RCL
13 13
65 ×
43 RCL
11 11
65 ×
43 RCL
24 24
95 =
44 SUM
19 19
22 INV
97 DSZ
09 09
68 NOP
43 RCL
22 22
44 SUM
10 10
97 DSZ
08 08
42 STO
06 6
```

```
42  STO
24   24
02   2
42  STO
08   08
61  GTO
42  STO
76  LBL
68  NOP
89   π
55   ÷
02   2
95   =
32  X↕T
43  RCL
21   21
42  STO
10   10
67   EQ
59  INT
97  DSZ
07   07
58  FIX
76  LBL
59  INT
98  ADV
04   4
04   4
69  OP
04   04
43  RCL
37   37
69  OP
06   06
04   4
05   5
69  OP
04   04
43  RCL
36   36
69  OP
06   06
03   3
06   6
01   1
05   5
69  OP
04   04
43  RCL
19   19
65   ×
43  RCL
44   44
65   ×
43  RCL
22   22
55   ÷
07   7
55   ÷
89   π
65   ×
43  RCL
35   35
95   =
69  OP
06   06
92  RTN
76  LBL
14   D
43  RCL
42   42
42  STO
00   00
75   -
01   1
95   =
35  1/X
65   ×
53   (
43  RCL
39   39
75   -
43  RCL
38   38
42  STO
36   36
54   )
95   =
42  STO
39   39
53   (
43  RCL
41   41
75   -
43  RCL
40   40
42  STO
37   37
54   )
55   ÷
53   (
43  RCL
06   06
75   -
01   1
54   )
95   =
42  STO
41   41
76  LBL
43  RCL
71  SBR
12   B
61  GTO
44  SUM
76  LBL
24  CE
71  SBR
22  INV
76  LBL
44  SUM
71  SBR
13   C
22  INV
97  DSZ
00   00
34  √X
43  RCL
39   39
44  SUM
36   36
61  GTO
24  CE
76  LBL
34  √X
22  INV
97  DSZ
06   06
35  1/X
43  RCL
42   42
42  STO
00   00
43  RCL
38   38
42  STO
36   36
43  RCL
41   41
44  SUM
37   37
61  GTO
43  RCL
76  LBL
35  1/X
91  R/S
76  LBL
12   B
70  RAD
29  CP
43  RCL
32   32
75   -
43  RCL
34   34
95   =
42  STO
23   23
94  +/-
22  INV
77   GE
23  LNX
89   π
55   ÷
02   2
95   =
42  STO
21   21
61  GTO
33  X²
```

```
356  76 LBL
357  23 LNX
358  43 RCL
359  37  37
360  55  ÷
361  43 RCL
362  23  23
363  95  =
364  22 INV
365  30 TAN
366  42 STO
367  21  21
368  76 LBL
369  33 X^2
370  43 RCL
371  37  37
372  55  ÷
373  53  (
374  43 RCL
375  33  33
376  75  -
377  43 RCL
378  34  34
379  54  )
380  95  =
381  22 INV
382  30 TAN
383  42 STO
384  23  23
385  76 LBL
386  22 INV
387  43 RCL
388  30  30
389  75  -
390  43 RCL
391  36  36
392  95  =
393  55  ÷
394  43 RCL
395  37  37
396  95  =
397  22 INV
398  30 TAN
399  42 STO
400  01  01
401  43 RCL
402  31  31
403  75  -
404  43 RCL
405  36  36
406  95  =
407  55  ÷
408  43 RCL
409  37  37
410  95  =
411  22 INV
412  30 TAN
413  42 STO
414  02  02
415  87 IFF
416  01  01
417  13  C
418  92 RTN
419  76 LBL
420  17 B'
421  86 STF
422  01  01
423  61 GTO
424  12  B
425  76 LBL
426  19 D'
427  22 INV
428  86 STF
429  01  01
430  43 RCL
431  43  43
432  42 STO
433  06  06
434  61 GTO
435  14  D
```

Sky Component Program: Clear Sky

```
000  76 LBL
001  16 A'
002  42 STO
003  15  15
004  39 COS
005  42 STO
006  16  16
007  33 X^2
008  53  (
009  42 STO
010  17  17
011  65  ×
012  43 RCL
013  12  12
014  85  +
015  43 RCL
016  13  13
017  54  )
018  42 STO
019  18  18
020  53  (
021  53  (
022  43 RCL
023  26  26
024  65  ×
025  43 RCL
026  20  20
027  65  ×
028  53  (
029  43 RCL
030  15  15
031  85  +
032  43 RCL
033  27  27
034  54  )
035  39 COS
036  85  +
037  43 RCL
038  11  11
039  65  ×
040  43 RCL
041  28  28
042  65  ×
043  43 RCL
044  16  16
045  54  )
046  55  ÷
047  43 RCL
048  18  18
049  34 ΓX
050  54  )
051  42 STO
052  29  29
053  53  (
054  53  (
055  53  (
056  53  (
057  43 RCL
058  13  13
059  85  +
060  43 RCL
061  17  17
062  65  ×
063  43 RCL
064  14  14
065  54  )
066  55  ÷
067  43 RCL
068  18  18
069  54  )
070  45 Y^X
071  01  1
072  93  .
073  05  5
074  85  +
075  01  1
076  54  )
077  65  ×
078  53  (
079  93  .
080  09  9
081  01  1
082  85  +
083  01  1
084  00  0
085  65  ×
086  53  (
087  03  3
088  94 +/-
089  65  ×
090  43 RCL
091  29  29
092  22 INV
093  39 COS
094  54  )
095  22 INV
096  23 LNX
097  85  +
098  93  .
099  04  4
100  05  5
101  65  ×
102  43 RCL
103  29  29
104  33 X^2
105  54  )
106  65  ×
107  53  (
```

```
108  01  1
109  75  -
110  53  (
111  93  .
112  03  3
113  02  2
114  94  +/-
115  65  ×
116  43  RCL
117  18  18
118  34  √X
119  55  ÷
120  43  RCL
121  11  11
122  55  ÷
123  43  RCL
124  16  16
125  54  )
126  22  INV
127  23  LNX
128  54  )
129  65  ×
130  43  RCL
131  17  17
132  65  ×
133  43  RCL
134  16  16
135  55  ÷
136  53  (
137  43  RCL
138  18  18
139  45  Y^X
140  02  2
141  93  .
142  05  5
143  54  )
144  54  )
145  92  RTN
146  76  LBL
147  28  LOG
148  65  ×
149  89  π
150  55  ÷
151  01  1
152  08  8
153  00  0
154  95  =
155  92  RTN
156  76  LBL
157  11  A
158  65  ×
159  01  1
160  93  .
161  00  0
162  01  1
163  08  8
164  95  =
165  42  STO
166  44  44
167  91  R/S
168  42  STO
169  45  45
170  91  R/S
171  42  STO
172  46  46
173  71  SBR
174  28  LOG
175  42  STO
176  27  27
177  43  RCL
178  46  46
179  91  R/S
180  42  STO
181  47  47
182  71  SBR
183  28  LOG
184  94  +/-
185  44  SUM
186  27  27
187  43  RCL
188  47  47
189  91  R/S
190  42  STO
191  48  48
192  75  -
193  09  9
194  00  0
195  95  =
196  94  +/-
197  71  SBR
198  28  LOG
199  42  STO
200  25  25
201  70  RAD
202  38  SIN
203  42  STO
204  20  20
205  43  RCL
206  25  25
207  39  COS
208  42  STO
209  28  28
210  43  RCL
211  48  48
212  91  R/S
213  76  LBL
214  13  C
215  43  RCL
216  23  23
217  42  STO
218  10  10
219  94  +/-
220  85  +
221  43  RCL
222  21  21
223  95  =
224  55  ÷
225  04  4
226  42  STO
227  09  09
228  95  =
229  42  STO
230  22  22
231  00  0
232  42  STO
233  19  19
234  42  STO
235  08  08
236  02  2
237  42  STO
238  07  07
239  76  LBL
240  58  FIX
241  03  3
242  42  STO
243  24  24
244  76  LBL
245  42  STO
246  02  2
247  22  INV
248  44  SUM
249  24  24
250  43  RCL
251  10  10
252  39  COS
253  42  STO
254  11  11
255  33  X²
256  42  STO
257  12  12
258  42  STO
259  14  14
260  43  RCL
261  10  10
262  38  SIN
263  42  STO
264  26  26
265  33  X²
266  42  STO
267  13  13
268  94  +/-
269  44  SUM
270  14  14
271  04  4
272  36  PGM
273  09  09
274  13  C
275  36  PGM
276  09  09
277  14  D
278  65  ×
279  43  RCL
280  13  13
281  65  ×
282  43  RCL
283  11  11
284  65  ×
285  43  RCL
286  24  24
287  95  =
288  44  SUM
289  19  19
290  22  INV
291  97  DSZ
292  09  09
293  68  NOP
294  43  RCL
295  22  22
296  44  SUM
297  10  10
298  97  DSZ
299  08  08
```

```
300  42 STO
301  06   6
302  42 STO
303  24  24
304  02   2
305  42 STO
306  08  08
307  61 GTO
308  42 STO
309  76 LBL
310  68 NOP
311  89   π
312  55   ÷
313  02   2
314  95   =
315  32 X:T
316  43 RCL
317  21  21
318  42 STO
319  10  10
320  67  EQ
321  59 INT
322  97 DSZ
323  07  07
324  58 FIX
325  76 LBL
326  59 INT
327  98 ADV
328  04   4
329  04   4
330  69 OP
331  04  04
332  43 RCL
333  37  37
334  69 OP
335  06  06
336  04   4
337  05   5
338  69 OP
339  04  04
340  43 RCL
341  36  36
342  69 OP
343  06  06
344  03   3
345  06   6
346  01   1
347  05   5
348  69 OP
349  04  04
350  43 RCL
351  19  19
352  65   ×
353  43 RCL
354  44  44
355  65   ×
356  43 RCL
357  22  22
358  55   ÷
359  03   3
360  55   ÷
361  43 RCL
362  45  45
363  65   ×
364  43 RCL
365  35  35
366  95   =
367  69 OP
368  06  06
369  92 RTN
370  76 LBL
371  19 D'
372  22 INV
373  86 STF
374  01  01
375  43 RCL
376  43  43
377  42 STO
378  06  06
379  76 LBL
380  14   D
381  43 RCL
382  42  42
383  42 STO
384  00  00
385  75   -
386  01   1
387  95   =
388  35 1/X
389  65   ×
390  53   (
391  43 RCL
392  39  39
393  75   -
394  43 RCL
395  38  38
396  42 STO
397  36  36
398  54   )
399  95   =
400  42 STO
401  39  39
402  53   (
403  43 RCL
404  41  41
405  75   -
406  43 RCL
407  40  40
408  42 STO
409  37  37
410  54   )
411  55   ÷
412  53   (
413  43 RCL
414  06  06
415  75   -
416  01   1
417  54   )
418  95   =
419  42 STO
420  41  41
421  76 LBL
422  43 RCL
423  71 SBR
424  12   B
425  61 GTO
426  44 SUM
427  76 LBL
428  24 CE
429  71 SBR
430  22 INV
431  76 LBL
432  44 SUM
433  71 SBR
434  13   C
435  22 INV
436  97 DSZ
437  00  00
438  34 ΓX
439  43 RCL
440  39  39
441  44 SUM
442  36  36
443  61 GTO
444  24 CE
445  76 LBL
446  34 ΓX
447  22 INV
448  97 DSZ
449  06  06
450  35 1/X
451  43 RCL
452  42  42
453  42 STO
454  00  00
455  43 RCL
456  38  38
457  42 STO
458  36  36
459  43 RCL
460  41  41
461  44 SUM
462  37  37
463  61 GTO
464  43 RCL
465  76 LBL
466  35 1/X
467  91 R/S
468  76 LBL
469  17 B'
470  86 STF
471  01  01
472  76 LBL
473  12   B
474  29 CP
475  43 RCL
476  32  32
477  75   -
478  43 RCL
479  34  34
480  95   =
481  42 STO
482  23  23
483  94 +/-
484  22 INV
485  77  GE
486  23 LNX
487  89   π
488  55   ÷
489  02   2
490  95   =
491  42 STO
```

```
21  21
61  GTO
33  X²
76  LBL
23  LNX
43  RCL
37  37
55  ÷
43  RCL
23  23
95  =
22  INV
30  TAN
42  STO
21  21
76  LBL
33  X²
43  RCL
37  37
55  ÷
53  (
43  RCL
33  33
75  -
43  RCL
34  34
54  )
95  =
22  INV
30  TAN
42  STO
23  23
76  LBL
22  INV
43  RCL
30  30
75  -
43  RCL
36  36
95  =
55  ÷
43  RCL
37  37
95  =
22  INV
30  TAN
42  STO
01  01
43  RCL
31  31
75  -
43  RCL
36  36
95  =
55  ÷
43  RCL
37  37
95  =
22  INV
30  TAN
42  STO
02  02
87  IFF
01  01
13  C
92  RTN
```

Reflected Component Program

```
76  LBL
33  X²
43  RCL
26  26
55  ÷
02  2
95  =
42  STO
09  09
92  RTN
76  LBL
44  SUM
43  RCL
29  29
85  +
43  RCL
28  28
95  =
55  ÷
43  RCL
28  28
95  =
49  PRD
09  09
92  RTN
76  LBL
11  A
71  SBR
33  X²
04  4
42  STO
12  12
01  1
42  STO
02  02
61  GTO
23  LNX
76  LBL
12  B
71  SBR
33  X²
43  RCL
28  28
42  STO
02  02
00  0
42  STO
12  12
61  GTO
23  LNX
76  LBL
13  C
71  SBR
33  X²
71  SBR
44  SUM
01  1
42  STO
```

```
02  02
04  4
42  STO
12  12
61  GTO
23  LNX
76  LBL
14  D
71  SBR
33  X²
71  SBR
44  SUM
43  RCL
28  28
42  STO
02  02
00  0
42  STO
12  12
76  LBL
23  LNX
43  RCL
27  27
42  STO
01  01
43  RCL
15  15
65  ×
43  RCL
16  16
95  =
42  STO
10  10
65  ×
53  (
43  RCL
24  24
75  -
43  RCL
25  25
54  )
95  =
42  STO
03  03
43  RCL
17  17
65  ×
43  RCL
18  18
95  =
42  STO
05  05
42  STO
07  07
42  STO
08  08
44  SUM
08  08
65  ×
43  RCL
23  23
95  =
42  STO
04  04
```

```
44 SUM
03 03
43 RCL
05 05
65 ×
43 RCL
22 22
95 =
42 STO
06 06
44 SUM
03 03
43 RCL
18 18
65 ×
43 RCL
19 19
65 ×
02 2
65 ×
42 STO
13 13
44 SUM
08 08
43 RCL
25 25
95 =
44 SUM
03 03
43 RCL
17 17
65 ×
43 RCL
19 19
65 ×
44 SUM
13 13
02 2
65 ×
44 SUM
08 08
43 RCL
25 25
95 =
44 SUM
03 03
43 RCL
20 20
85 +
43 RCL
16 16
55 ÷
02 2
95 =
55 ÷
43 RCL
19 19
95 =
42 STO
14 14
65 ×
43 RCL
13 13
65 ×
44 SUM
05 05
43 RCL
25 25
85 +
43 RCL
04 04
95 =
55 ÷
43 RCL
05 05
95 =
49 PRD
01 01
01 1
75 -
43 RCL
14 14
95 =
65 ×
43 RCL
13 13
65 ×
44 SUM
07 07
43 RCL
25 25
85 +
43 RCL
06 06
95 =
55 ÷
43 RCL
07 07
95 =
65 ×
43 RCL
09 09
85 +
43 RCL
01 01
95 =
65 ×
43 RCL
21 21
65 ×
43 RCL
10 10
55 ÷
53 (
43 RCL
08 08
75 -
43 RCL
03 03
54 )
95 =
65 ×
43 RCL
02 02
95 =
42 STO
11 11
98 ADV
03 3
05 5
01 1
05 5
69 OP
04 04
43 RCL
11 11
58 FIX
40 IND
12 12
69 OP
06 06
22 INV
58 FIX
98 ADV
91 R/S
```

Cityscapes

Chapter Three

Daylight in Manhattan

In major cities, sun and light only reach the upper levels of the tallest buildings

If natural light replaces the use of electricity in large buildings, energy consumption will be greatly reduced. But many of our cities are so congested that sun and light only reach the upper levels of the tallest structures. Lower stories, low-rise buildings, and the streets are shaded throughout the day. Now, New York City has moved to reverse this dilemma with a newly proposed set of zoning regulations for midtown Manhattan.

The proposed regulations consider both the quantity and quality of natural light reaching the city's streets to be of primary importance. The creators of the proposal feel that the new laws could become the pivotal reference for new zoning ordinances in other American cities. At the very least, they are certain that the regulations will have two major effects on Manhattan: buildings will consume less energy and the quality of pedestrian street life will improve.

Zoning laws

New York's first set of zoning laws reflected a concern with daylighting

The use of zoning as a means of regulating natural light is not foreign to New York. The city's first set of zoning laws reflected a concern with daylighting. In 1915, the Equitable Life Insurance Company opened the doors of its new headquarters at 120 Broadway and was immediately hit with a wave of protest. The 40-story skyscraper was the tallest building in the city, and it created a canyon of shadows along nearby buildings and streets. Property owners in the area were furious, and their anger at the loss of natural light led to America's first comprehensive attempt at municipal regulation of building development. The ensuing 1916 New York Zoning Regulations literally prescribed the forms now synonymous with midtown Manhattan: the wide avenues, the uniform street wall, the classic skyline punctuated by individual towers such as the Chrysler, RCA, and Empire State buildings.

Buildings were prohibited from impinging on a prescribed "sky exposure plane"

To prevent the canyon effect, the regulations divided Manhattan into a series of districts, each with a limit on the height of buildings and their street walls. These limits were in direct proportion to the width of the streets that fronted the buildings: for each foot the structure set back from the street, it could rise an additional number of feet. Buildings were prohibited from impinging on a prescribed "sky exposure plane," an angle measured from the center of the fronting street to the peak of the building. The rise-to-setback ratio and the sky angle plane varied from district to district depending on street width, existing land use, and property values. The two most common districts had rise-to-setback ratios of 2.5:1 and 3:1. These corresponded with sky angles of 68.3° and 71.6°, or an average of 70°. Buildings were permitted to have towers of unlimited height, but these had to be set back a mandatory distance from the street, and could cover no more than 25 percent of the overall lot area.

The City Planning Commission adopted a completely new type of zoning regulation

Over the years, variances and the changing needs of the city decreased the effectiveness of the 1916 ordinances. In 1961, partially as a final reaction to the "wedding cake" shape of the skyline—created by less imaginative builders trying to comply with the 1916 laws—the City Planning Commission adopted a completely new type of zoning regulation A "sky exposure plane" replaced the height districts for governing setbacks. The permissible tower area of a building was increased to 40 percent of the overall low area, and a new concept, the Floor Area Ratio (FAR) was introduced to control a building's bulk. In the initial design, the floor area of a proposed building could be no greater than 15 times the overall lot area. However, by providing outdoor amenities such as plazas, builders could get the city to raise the allowable FAR up to 18.

Developers and designers soon found that the new ordinances favored tall, monolithic structures: towers and slabs. Although the effect of these buildings on daylighting levels was not drastic, city planners quickly began to feel that the isolated structures were destroying the existing character of the city. The Planning Commission responded to this dilemma by expanding the Special Permits section of the zoning regulations to include interior amenities. Now, if a new proposal included items such as arcades or mid-block passageways, its FAR could go as high as 21.6. This "incentive zoning" was an attempt to promote buildings which, once again, assumed their traditional role of defining the public space of the street.

Unfortunately, the discretionary review process that accompanied incentive zoning was inherently subjective and frequently disregarded the traditional need for light and air in the street. Residents and pedestrians in midtown soon discovered that the slabs and towers produced by incentive zoning did not make good neighbors; the amenities at the base of the buildings did not make up for the height and bulk at the top. Ironically, many of the Special Permit buildings, which rise up straight from the street edge, have no more bulk than earlier buildings in the city; the placement of the bulk, however, drastically reduces daylight and street space. Joanna Battaglia, head of a community planning board, eloquently expresses the frustrations that many New Yorkers feel: "We are having unwanted amenities foisted on us at the expense of the best amenities of all—light and air."

Many of the Special Permit buildings have no more bulk than earlier buildings but reduce daylight and street space

Providing for light and air

In 1979, the city authorized a study to revise the existing regulations. Much of the work for the recently published report was done by Kwartler/Jones, Architects, and Davis, Brody & Associates. Their Height and Bulk Proposal returns to the spirit of the 1916 regulations. Once again, admission of light and air to the street are primary zoning considerations, but the new ordinances specifically avoid dictating any particular building shape or form. Instead, the new proposal is based on a standard of daylight and openness derived from the actual condi-

In 1979, the city authorized a study to revise the existing regulations

tions that exist in midtown Manhattan—many of which are the direct result of the 1916 regulations. The consultants made considerable efforts to ensure that today's architects will have as much design freedom as possible in creating buildings that comply with these conditions.

The goal of envelope-based zoning is to ensure year-long solar access to every building within its jurisdiction

Shortly after it received the city's contract, Kwartler/Jones hired Harvey Bryan as a consultant. Bryan, currently a professor in the Department of Architecture at MIT, had done a good deal of lighting research and was a staunch advocate of urban daylighting. Initially, the team was asked to explore the merits of the "solar envelope" urban design concepts developed by Ralph Knowles. The goal of envelope-based zoning is to ensure year-long solar access to every building within its jurisdiction. Very quickly, the consultants came to the conclusion that Knowles' concept was technically ill-suited to New York City. "Most of the buildings in midtown are internal gain buildings: cooling load dominated. They don't need solar access; in fact, most of them need less direct sunlight, not more."

Beyond a certain point, however, too much shading from surrounding buildings becomes a liability to life on the streets. The consultants needed a method of determining the level of daylight, sunlight, and shade that was optimal for building energy efficiency and pedestrian life. They began background research to discover historical examples of urban zoning based on daylighting. "Surprisingly, New York's own 1916 regulations were one of the best references," says Bryan. "A lot of studies on urban daylighting had been done in Europe since that time, but little had ever been applied."

The daylight map

Too much shading from surrounding buildings becomes a liability to life on the streets

The consultants also began assessing Manhattan's existing daylighting problems. They went throughout the midtown area, measuring shadows and taking photographs. To help in their work, they developed a modified version of the Waldram diagram, a graphic tool developed in England during the 1930's. By using this diagram to analyze representative buildings and city blocks, the consultants were able to develop a "Daylight Map" of midtown Manhattan. This map was then used to identify the minimum and average expected levels of daylight in midtown resulting from 64 years of zoning regulations.

In June 1980, the consultants handed the Planning Commission a performance based zoning proposal. According to the proposal, every future building project would be assessed by four criteria: daylighting, street wall height, street wall length, and reflectivity. To receive clearance, the proposed building had to score at least 85 points out of a 100-point evaluation. Street wall height and wall length were included as criteria to ensure that new projects would respond favorably to the existing physical characteristics of their surroundings. Reflectivity was an optional criterion. Building designs were rewarded for using light-colored materials on the outside. Daylighting, however, was by far the most important factor in the proposed ordinance. Buildings were judged by the amount of natural light permitted to reach the streets surrounding them, according to a simple, accurate method developed by the consultants (see daylighting test).

Buildings were judged by the amount of natural light permitted to reach the streets surrounding them

The Planning Commission had a mixed response. It wanted to increase the quantity and quality of light reaching the street, but it also felt that the proposal, as written, was too confusing for the average architect, builder, or developer. The New York chapter of the American Institute of Architects supported the consultant's approach, but the Real Estate Board of New York, worried about the effect that the regulations would have on future development, told the commission that the new concept was "excessively cumbersome." The consultants rewrote the initial draft, deleting much of the technical data and simplifying the performance requirements. The Planning Commission endorsed the revision but still voiced the concern that it was "too cumbersome." The commissioners countered with a two-tiered proposal, which contained a modified version of the consultant's performance approach and an even simpler, although architecturally more restrictive, prescriptive approach. To receive zoning approval, a proposed project would have to fulfill the requirements of only one, not both, of the approaches.

Still, both parts of the city's final report, which was issued during the summer of 1981, are based on daylighting. The performance approach is similar to the consultants' original proposal, with minor changes in scoring the four tests. The prescriptive approach is similar in theory to the sky-angle planes of the 1916 ordinance. The approach uses "sky exposure curves," which define the allowable shape and bulk of newly proposed projects.

Street level daylight is an amenity needed in every city

Street level daylight is an amenity needed in every city. Just as the 1916 regulations became the prototype zoning laws for other U.S. cities, the concepts behind the new revisions could be adapted to other situations. Many Sun Belt cities are expanding rapidly. If they aren't careful, they will soon have the same problems that Manhattan has. A rigorous set of zoning regulations based on daylighting would greatly protect the basic amenities of these cities.

Daylighting Test

The Daylight Evaluation Chart is a modified version of the Waldram diagram

The daylighting performance test attempts to quantify the amount of natural light that will reach the street after a proposed project has been built. Light to the street is measured on a Daylight Evaluation Chart, which is a modified version of the Waldram diagram developed in England during the 1930's to aid in daylighting analysis. The modified diagram is a graphic representation of a three-dimensional view of a building; it presents a slightly distorted version of a proposed project as it would be seen by a pedestrian standing at either end of the street fronting the building. The variable grid prestamped on the diagram helps quantify lighting levels. Each square represents two vertical degrees of sky and 25 feet of lot frontage at the street line, so that for a site occupying 250 feet of street length, a full 100 available daylight squares would become the basis for evaluating daylight performance.

The variable grid prestamped on the diagram helps quantify lighting levels

The consultants, realizing that "below the range of prevailing street walls in midtown, there is little expectation of seeing the sky," distorted the grid to give greater value to daylight coming from higher in the sky. They determined that the average street wall height presently found in Manhattan occurred at approximately a 70° angle above the ground measured from the center of the fronting street. Therefore, they divided their diagram at the 70° point. Above this level, the grid is divided into 100 blocks whose size varies in relation to the "quality" of the light coming through these areas. Unobstructed blocks above the 70° line are multiplied by their weighted value, totaled, and subtracted from 100. Blocks below the 70° line are similarly multiplied, but their weighted values are less. These values are also totaled and added to the original figure; the sum of the two figures represents the building's final daylighting score.

DAYLIGHT EVALUATION CHART; 60 FOOT STREET (COPYRIGHT, 1981)

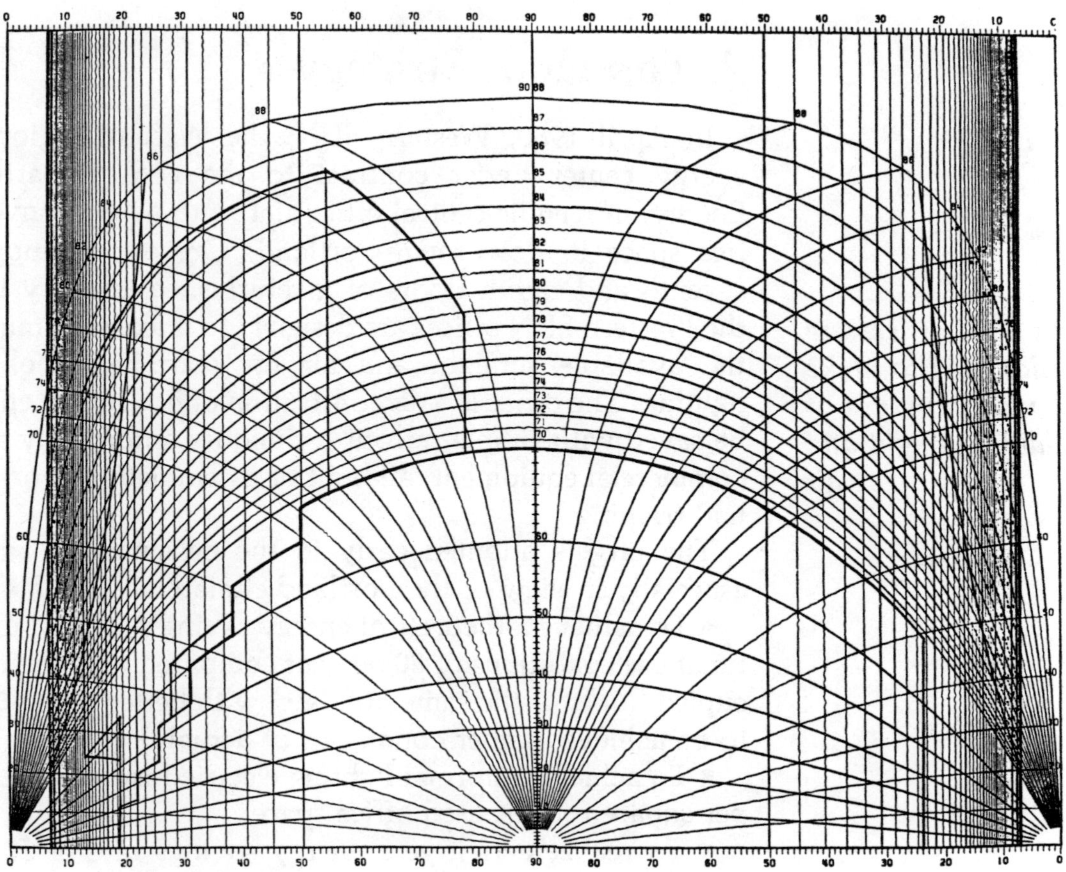

Citicorp Center Score

Available Daylight Squares	128
Squares Blocked Times Value	−40.23
Squares above 70° blocked	13.8
Squares below 70° open	−16
Total blockage	−42.3

Daylight Score

$$\frac{128 - 42.3}{128} = 67\%$$

Using reflectivity to increase score R_v (reflective rating) of the building:

aluminum = .85 (60% of building surface)

solar ban glass = .38 (40% of building surface)

Reflectance

.85 × 60% + .38 × 40% = .662

Then determine building face orientation:

36 squares face S, SW

4.23 squares face E, SE

orientation value S, SW = .93

orientation value E, SE = .72

Formula:

reflectivity=(R_v − .15) × (O_v × number of squares)

Reflectivity for S, SW=(.662 − .15) × .93 × 36 squares=17.14

Reflectivity for E, SE (.662 − .15) × .72 × 4.23 squares=1.56

Total 18.70

Now apply reflectivity adjustment to original score

128 − 42.3 + 18.7 ÷ 128=81.5%

Therefore, light materials increase the score for the building 14.5%

A Top-Floor Strategy

Western SUN offices are designed to make maximum use of available sunlight and air

In April 1980, Western SUN, the Western regional solar energy center under contract to the U.S. Department of Energy, moved its central office staff into newly converted office space in downtown Portland, Oregon. Occupying the seventh and eighth floors of a refurbished 10-story building, the Western SUN offices are designed to make maximum use of the available sunlight and air for warming, cooling, and lighting. The office is a demonstration of how much energy can be saved with simple retrofit techniques and readily available commercial equipment, all built or assembled by workers who had training.

The targets adopted early in the design were ambitious, especially in view of the Portland climate:

- To achieve a theoretical energy saving potential under optimal use conditions of 80 percent compared with the other occupied floors of the same building, with better than 60 percent to be achieved under conditions of actual use;
- To achieve a normal lighting electric load of about 1 watt per square foot of used office space;
- To realize a first-year energy savings rate of $2,500 per year per floor (or $5,000 for the two floors) at the Portland commercial cost of three cents per kilowatt hour; and
- To accomplish a payback of five years or less for extra costs incurred as they relate specifically to energy conservation and to the use of natural energy resources.

Heating

Since bright sun produces glare, staffers will close the venetian blinds to block direct gain

Heating can be achieved in south-facing offices by admitting direct solar gain on sunny days, with circulation of the warmed air promoted by open doors in walled-in offices and by an open floor plan with movable walls in the bulk of the office space.

Since bright sun produces glare, staffers will close the venetian blinds enough to block the direct gain. This facilitates the indirect mode of solar heating in which the return air to the fan room is drawn past the venetian blinds (**Figure 23**). The standard dark brown metal blinds, coupled with the addition of inner double-hung windows and a low-speed fan to slow down the velocity of return air, leads to venetian blind air collectors of surprising efficiency (the design estimate is better than 40 percent). Of course, this method allows the solar-heated air to

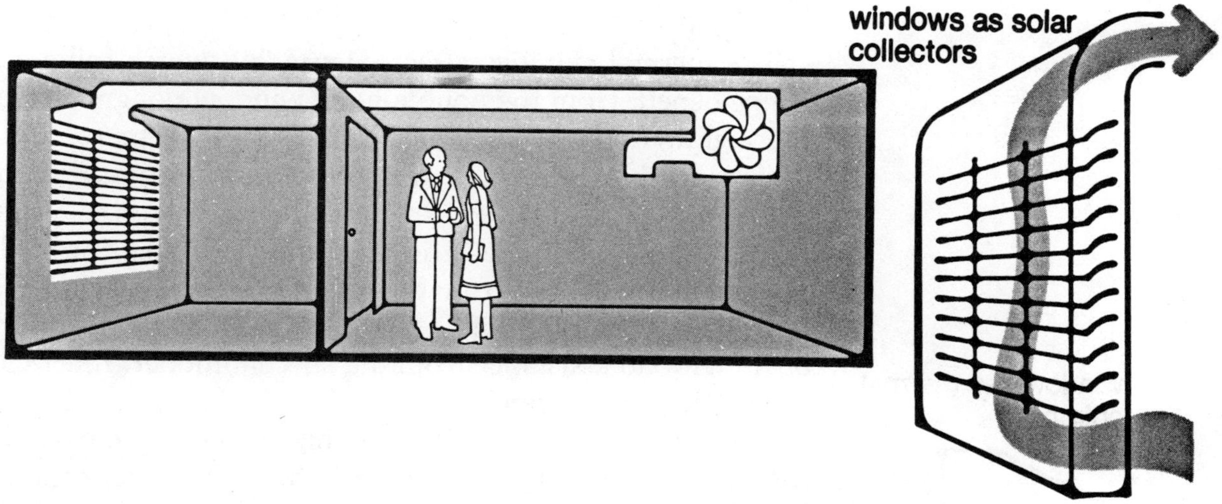

Fig. 23 — Lowering the dark venetian blinds to prevent direct sun from overheating the south-facing offices causes the blinds to become efficient solar energy collectors. By drawing the return air to the fan room past the solar-heated blinds, the warmed air is redistributed through the normal ventilation system to all of the office spaces. On sunny days the furnace will not be needed.

be distributed to all of the space rather than just in the collecting east-, south-, and west-facing perimeter offices.

At the onset of the study, in mid-November 1980, on sunny days, with the outside temperature in the forties, the interior offices stabilized at about 70°F with no backup heating. The slow air movement in the low-fan-speed solar mode produces a stuffy feeling, so some windows were invariably opened a bit to provide a little fresh air.

With both inner and outer sash windows closed and all lights located in the working environment rather than in the ceiling, the interior heat gains from lights and human bodies, along with heat loss reductions by double-glazing, appeared to keep the spaces warmed nearly to the federally mandated minimum of 65°F on chilly, rainy winter days without the use of the backup electric duct heaters, even for early morning warm-ups.

In short, the Western SUN offices could be self-heated in virtually all winter weather conditions, as designed. Nevertheless, should backup heating be called for, the system could be put in a normal, high-speed-fan override setting with the automatic opening of properly sized inner office return air dampers and the automatic closing of dampers from the window-chimney ducts.

At the onset of the study, the interior offices stabilized at about 70°F with no backup heating

Cooling

The thermal chimney effect (**Figure 24**), which follows as a fringe benefit from the double-hung window design, works on the eighth floor, where the airflow from north to south is unimpeded by fixed walls. In this case, the shady-side cooling of the building air drawn in by the solar-energized venetian blinds, along with the efficient venting of the warmed air around the blinds and from the offices, allowed the eighth floor to remain at about 74°F (with the outside air temperature at 88°F) with no assistance from the air conditioner. Inner closed offices, however, were stuffy, so that the fans in the fan room were normally operated in their high-speed setting anyway, though with the air conditioner compressor interlocked off. Merely moving the air in and out of the inner offices provided sufficient comfort.

Fig. 24 — Since there are no shading overhangs on the Western SUN building, in the summer the east-, south-, and west-facing venetian blinds will be drawn to keep the space cool. The sun will still heat the blinds, but opening the top of the outer sash window and the bottom of the inner window allows the heated blinds to act as thermal chimneys. These chimneys serve as solar-powered pumps to draw cool air from the north windows through the space, while venting the heated air to the outside. This air motion, plus the infusion of air from the night-cooled side of the building, provides for a natural cooling action, allowing the air conditioner to remain off generally until the outside temperature exceeds about 90°F.

The seventh floor fared less well, for the wall of a library isolates the north from the south quadrants of that space, thus destroying the natural ventilating airflow. The occupants of the seventh floor were forced to use air conditioning much more frequently in warm weather. However, careful zoning of the air conditioned sectors, coupled with open windows in the early cooler hours of the day and the very low electric light energy levels, still led to a significant energy savings in comparison with the control floors of the building.

Lighting

The lighting was designed to be electric-light-assisted daylighting rather than the reverse. Ceiling acoustical panels with high-reflectance paint (sold as a standard item under the trade name Nubby Two, by Armstrong) softly and efficiently diffused the solar radiation, which was reflected upward by the slightly tilted venetian blinds.

An agreement with the building owner prevented painting or silvering the upper portions of the venetian blinds, but the lacquered metal blind surface provided a very satisfactory specular reflection of a sufficient amount of the sun's rays for good unassisted daylighting in perimeter offices.

When it is cloudy or rainy, the open blinds admit sufficient diffuse radiation for soft supplemental daylighting in all perimeter offices. Glass inner office walls, plus glass beginning above 65 inches between all offices, allows for maximum sharing of the daylighting in all conditions.

All electric lighting is beamed up to the same areas that receive reflected daylight by using Steelcase "ambient" light fixtures mounted over the Steelcase desk book shelves; or it is beamed up and down with site-built "wall washers" placed just below the glass portions of office dividing walls (each upper electric light is shared across two offices). In some cases, it is beamed directly onto the work surfaces with the Steelcase "task" lights below the same book shelves (**Figure 25**).

windows admitting sunlight and background light

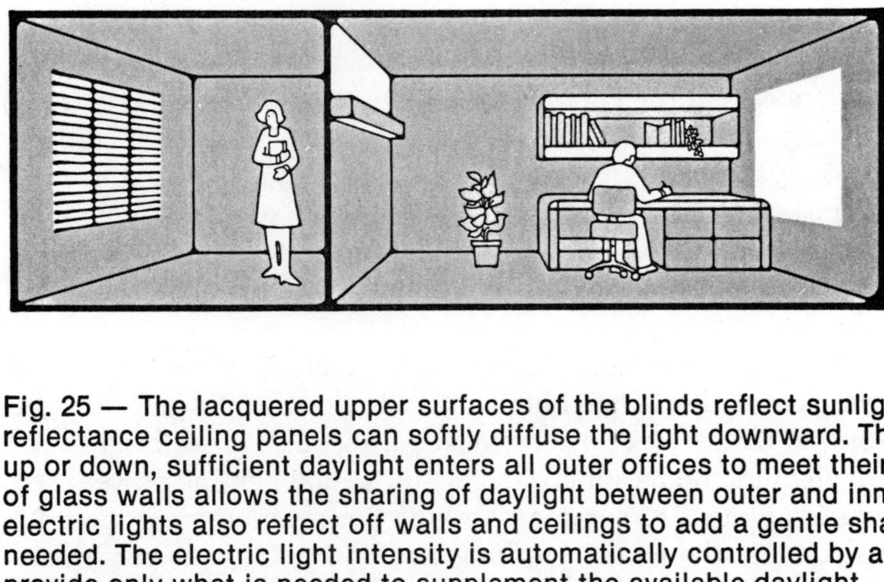

Fig. 25 — The lacquered upper surfaces of the blinds reflect sunlight upward, where high-reflectance ceiling panels can softly diffuse the light downward. Thus, with the blinds either up or down, sufficient daylight enters all outer offices to meet their needs. In addition, the use of glass walls allows the sharing of daylight between outer and inner offices. The mounted electric lights also reflect off walls and ceilings to add a gentle shadow-free contribution as needed. The electric light intensity is automatically controlled by an electric dimmer circuit to provide only what is needed to supplement the available daylight.

The lighting scheme provides a contrast-free visual environment and a shadow-free condition

This lighting scheme allows for automatic dimming of the ambient lights in direct proportion to the available daylighting. It also provides a contrast-free visual environment and a shadow-free condition that does not change perceptibly in light level or quality as the electric lighting and daylighting ratios vary.

Table 1 describes the installed electric lighting on the eighth floor. With all fixtures operating at once and no dimming (e.g., a nighttime condition), the total wattage in the 4,200-square-foot working area is 7,030, or 1.67 watts per square foot. While the agreement with the building owner prevents the use of reflecting light shelves and other daylight-enhancing mechanisms, this value would still seem to violate the 1 watt per square foot design goal. But this is not how the system operates in practice.

Table 1: Installed electric lighting—Western SUN floor 8

Type of lighting	Number of fixtures	Number of bulbs in each fixture	Wattage of bulbs	Dimming
Ambient	13	4 (can be used with 2)	40	yes
Ambient	9	4 (can be used with 2)	30	yes
Wall washers	21	2	40	yes
Overhead	3	4 (can be used with 2)	40	no
High Intensity Discharge (HID)	4	1	250	no
Task	11	1	40	no
Task	9	1	30	no

First, the ambient light fixtures each carry four bulbs, but two may be turned off in each with no loss of lighting efficiency during beamed daylight hours; light levels under diffuse (cloudy) conditions are still satisfactory. Under this circumstance, again without dimming and with all remaining lights on at once, the electric lighting rate reduces to 4,370 watts, or 1.04 watts per square foot, which is about equal to the design parameter.

Second, experience demonstrates that all perimeter office ambient and wall-washing lights dim down to 15 percent light output (20 percent power levels) during all beamed (bright sun) daylighting hours, thus reducing the actual lighting power used during almost all summer working hours and during about 30 percent of the winter working hours to 0.75 watts per square foot with all four bulbs in the ambient fixtures operating, and

0.64 watts per square foot in the two-bulb positions. During dark, rainy days, the perimeter light dimming is to about a 75 percent power level, which yields 0.95 watts per square foot (two-bulb fixtures) to 1.4 watts per square foot (four-bulb fixtures) during about 35 percent of the annual working hours.

Averaged over a normalized annual workday, the energy used for electric lighting is well below 1 watt per square foot.

The energy used for electric lighting is well below one watt per square foot

Energy savings and costs

Electric monitoring was provided on the top five occupied floors of the building (all used for offices), with separate monitoring for each of the two Western SUN floors. Since all space comfort conditioning and lighting is provided by electricity, with heating and cooling units on each floor, this method measures floor-by-floor aggregated lighting and space conditioning energy use. Data collected from two months of electric monitoring is shown in **Table 2.**

Table 2: Measured energy use and savings—both Western SUN floors (1980)

Month	Average Energy Use of Office Floors 6, 9 & 10 (kWh)	Energy Used by Western SUN (kWh) Floor 7	Floor 8	Energy Saved by Western SUN (kWh) Floor 7	Floor 8	Totaled Savings @ 3¢/kWh
Sept.	9,951	5,228	3,319	4,723 (= 47%)	6,632 (= 67%)	$341
Oct.	9,375	3,423	2,762	5,943 (= 63%)	6,613 (= 70.5%)	$377

The average energy savings of the eighth floor amounted to 67 percent for the first month and 70.5 percent for the second. The kWh savings for this floor were about 6,600 for each month, which converts to a savings of about $200 per month at three cents per kWh, or an annual savings of about $2,400. The diminished performance of the seventh floor still produced an energy savings of 47 percent for the first month (when air conditioning was still used) and 63 percent for the second, suggesting that even under wasteful circumstances (which included a malfunctioning compressor that ran for days), the energy-conserving design resulted in significant real cost savings.

The energy-conserving design resulted in significant real cost savings

The combined two-month energy savings of both floors averaged about $360 per month, for an annual rate of about $4,300.

Costs and cost-payback are difficult to assess in terms of true add-on figures that result exclusively from energy-conserving and solar features. The net cost for the construction of the 8,400-square-foot conversion was $190,376, which was $112,406 above the building lease allowance of $77,970. But, for example, glass office walls are a common interior design feature. Is the cost of glass (higher than the cost-allowed stud and sheet-rock walls) charged to electric light energy conservation, or to solar, or just to nice architecture? Is the cost of inner double-hung windows (about $14,000) charged to double-glazing energy conservation, or to solar? The light dimmers cost $750 each (to control up to 40 40-watt bulbs each), or $3,000 for the four zones on each floor. Should that be charged against electric energy conservation or to solar (daylighting)?

The true extra expenses for special things, such as installing inner glazing, wiring in dimming ballasts, installing two-speed fans in the furnaces, and ducting the east, south, and west windows to the return air lines, probably doesn't exceed $60,000 of the total over-the-building allowance. At an annual savings rate of $4,300, this would be a 13.9-year straight-line payback. But Portland electric utility rates have risen to 3.9 cents per kWh, a 30 percent jump, thus reducing the straight-line projection to 10.7 years. Five years to pay back $60,000 may still prove out to be the case.

Daylight Control

Chapter Four

Control is a key element in daylight design, in terms of both direct solar radiation and sky brightness and to ensure a desired quality, as well as quantity, of light. The quantity of daylight illumination in a space is a function of the amount of daylight available, the site, the fenestration, and the room. The critical elements in daylight availability are the time of day, season, cloud cover, microclimate, and air pollution. The site must be considered in terms of orientation, latitude, landscape, and obstructions. Fenestration concerns include sun control, the framing and wall system, the sash, transmission characteristics, maintenance, wall location, and size. The geometry of the space, plus its surface reflectance, maintenance, furniture, and task location, are also critical.

More light does not necessarily mean better light. The qualitative aspects of daylight and the tradeoffs between illumination and thermal design objectives must be considered.

More light doesn't necessarily mean better light

There are three basic reasons to include daylight controls in any design strategy: to eliminate excessive illumination on interior surfaces, to eliminate high sky brightness or direct radiation, and to redistribute sunlight by diffusing it.

Orientation is critical. Even on the north facade, a clear glass window can see 5,000 footlamberts on a cloudy day; glare control is necessary. Controls on the south, east, and west are, of course, more critical because these elevations are exposed to direct radiation, which poses not only issues of illumination control but of thermal control. Ways must be found to respond to the rhythms of the sun, rejecting heat when it is unwanted and admitting it when desired. Control, then, is the essence of design.

Exterior controls

Fixed exterior controls are used to limit excess sun or sky light

Fixed exterior controls are used to limit excess sun or skylight. Overhangs eliminate the impact of direct sunlight and heat on glass during specified times of the year. Designs may be flat or sloping; sloping versions may be of shorter length but obstruct more of the sky view. They may be translucent or opaque. Translucent overhangs have the advantage of providing weather control and the transmission advantages of louvers.

Exterior vertical or horizontal louvers may be useful to control sun and heat

Exterior louvered overhangs allow the reflected sunlight to illuminate the visible surface of the control to the room. If the louvers are not opaque, they allow better utilization of the sunlight and skylight in the room and also allow the dissipation of heat. The disadvantages are that louvers allow rain to come in and cannot be used along narrow exterior corridors.

Exterior vertical or horizontal louvers may be useful. The vertical louvers lend themselves to controlling sun on the east and west facades of a building; the horizontal louvers are most effective on the south side. Vertical louvers may also be used to control sky brightness on the north side.

A diminutive louver can be applied to a window and not greatly obstruct the view; if the slats are fine enough, the eye will maintain clear vision by putting the image together as it does in pointillist paintings or with pictures screened for publication. Even a very small louver will shield the sun at critical times but allow a clear view if put across a vision strip. In Europe, rolling diminutive louvers are available.

Other exterior fixed controls include screens, trellises, and shrubbery

Other exterior fixed controls include screens, trellises, and shrubbery. Trellises in particular have been used architecturally to control light, although the design must be carefully considered.

Fixed exterior amplifying controls enhance the impact of the sun and skylight rather than diminish it. These include surfaces that either diffuse or reflect the light. For instance, white stones and concrete diffuse light; polished aluminum and mirror surfaces redirect it.

Overhangs with reflective surfaces underneath transmit maximum ground reflections. The surface may be painted white or may be of concrete. Light shelves are overhangs located below the window to reflect light up toward the ceiling inside. Parapet walls are another variation. These are effective in combination with a north-facing clerestory; they reflect the light from the southern sky and sun through the clerestory and onto the ceiling. Louvered overhangs can be designed as a control to both limit solar radiation from direct penetration into a space and to amplify its efects by reflecting it into the interior of the building. (See **Fig. 26**.)

Overhangs with reflective surfaces underneath transmit maximum ground reflections

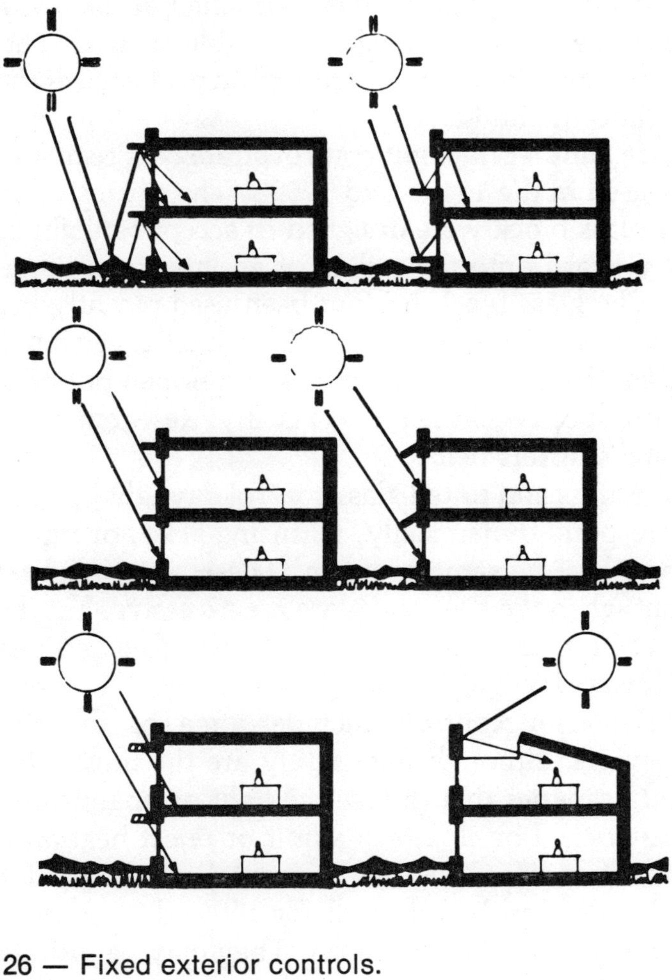

Fig. 26 — Fixed exterior controls.

Tracking mirrors are the most sophisticated exterior control systems

Variable exterior controls offer the greatest control potential by responding to changes in day and season. Such controls include window awnings or awnings attached to the edge of a fixed overhang. Louvers adjustable for tilt may also be used. On the east and west facades they may be drawn to the side; on the south side they may be raised and lowered according to the time of day and season. Tracking mirrors are the most sophisticated systems. They require dynamic control to either track or adjust incrementally to the apparent movement of the sun and to changing sky conditions. The cost of such systems must be considered in the context of the overall building performance and according to a life-cycle cost/benefit analysis.

Interior controls

Interior fixed controls include diffusers, such as translucent materials, to soften and diffuse sunlight. The material may be part of a shade device or may hang loosely between the natural light source and the space. Louvers, installed inside a window or between panes of glass, may be adjustable or fixed. They are generally less effective than exterior controls because they let more heat into the building.

The glass block has not been used recently because the interior surface gets very hot

Prisms are another daylight control strategy. Prismatic glass block was used in the 1940's and 1950's. The prisms on the inside of the glass block were designed to accept sunlight and to redirect it upward onto the ceiling of an interior space as diffuse light. The glass block has not been used recently because the interior surface gets very hot. A prismatic version of insulating glass has been experimentally developed but not commercially produced because it is too costly. Research may yield breakthroughs in this field.

Reflective glass and tinted glass control daylight but currently cannot respond dynamically. Diffusing glass, or milky patterned glass, has the same problem. These materials keep the sunlight out when the heat and glare are excessive, but they do the same when the heat or light may be desirable and result in generally lower daylight levels.

Variable interior controls include some of the simplest daylight devices

Variable interior controls include some of the simplest daylight devices, although they rarely are the most effective. Shades and draperies that diffuse the light or opaque versions that can be opened or closed to admit or reject heat and light are the most familiar. Variable horizontal or vertical louvers (venetian blinds) can be adjusted for tilt and/or removed in response to natural light conditions. They may be installed inside a window or between its panes of glass. Reflective-coated shades or blinds can be used to reject unwanted heat and light. A variation has a black side to absorb heat in cold weather and a reflective side to redirect it in summer. Properly adjusted louvers can be used to amplify the effects of daylight by reflecting and redirecting it. Research in this area promises new developments soon. Multiple shades and draperies may also be installed and manipulated to diffuse light under one condition and to reject it under others. A familiar example is the sheer curtain as a light diffuser with a heavy fabric drapery to draw over it to keep heat and light in or out of the space.

Properly adjusted louvers can be used to amplify the effects of daylight

Fixed interior amplifiers, or light shelves (**Fig. 27**) may be installed to take advantage of reflected light from below the horizontal or to reflect sunlight or sky brightness. The shelves operate similarly to exterior overhangs and may have either diffuse or specular reflective surfaces, although the specular surface can result in bright spots. Light-colored floors, ceilings, draperies, blinds, and furniture also amplify interior light.

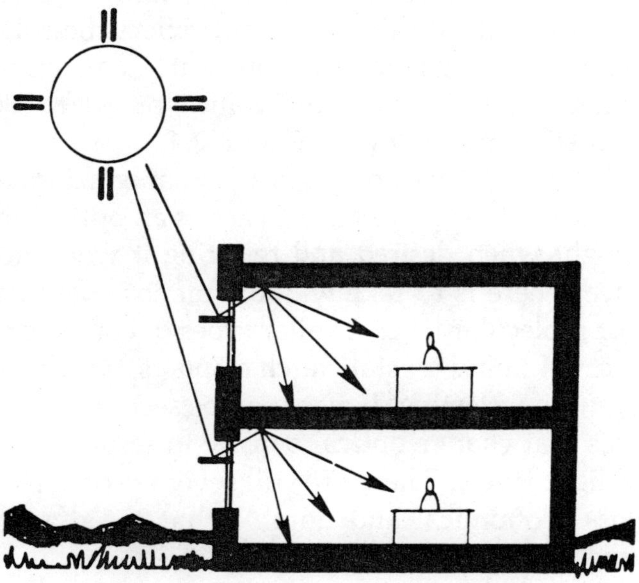

Fig. 27 — Light shelves.

The issue of manual versus automatic solar and glare controls is part of the design tradeoff. It seems likely that office occupants will close shades and blinds to reduce excessive heat gain or glare for thermal or visual comfort. It is not clear, however, whether they will operate these devices to save energy. The blind that is closed in the afternoon to prevent glare may not be reopened to admit daylight the next morning.

Glazing

A selective transmitting window will allow the visible light to pass through it

If interior and exterior shading devices are used to control excessive sun and glare, the question of glazing properties becomes important. Workers in a typical office have a view of the horizon or the sky in the immediate vicinity of the horizon. With standard overcast skies, the horizon is three times darker than the sky overhead. In clear sky conditions, however, the luminance distribution is inverted; and the horizon is brighter than the overhead sky. In urban areas, haze and air pollution can produce additional scattering of the light and additional glare. Because sky luminance at the horizon may be 500 to 3,000 footlamberts versus 25 to 75 footlamberts for a typical office, severe glare problems may result.

Some degree of light control in glazing is needed, but it is unlikely that transmissions of less than 50% would be desirable for daylighting. Heat absorbing and reflecting glass may also be desirable for such situations as sunlight reflected by water or from adjacent landscaping or buildings.

A selective transmitting window will allow the visible light to pass through it into the room but reflect the solar infrared portion back to the outside environment. Coatings of this type might have light transmission values of 60 to 80%, with shading coefficients as low as 0.3 or 0.4.

Glass can be used as an optical shutter, admitting or rejecting light as desired

A more speculative approach to solar glazing control involves coatings that cause glass to act as an optical shutter, admitting light when desired and rejecting it when not desired. The concept here is to take window management down to an atomic or molecular level. Similar types of coatings exist today in the form of liquid crystal watch displays, which switch from transparent to reflective. It also may be seen in the phototropic sunglasses that change optical density in response to changing ambient light levels. There are still severe problems in making such glass economical and durable, but the dynamic nature makes the concept very attractive.

Integration

Daylight can provide good visibility and amenity for indoor environments, but it will not reduce energy consumption unless people turn off the installed artificial lights. Daylight design may result in lower levels of installed lighting, but its greatest energy contribution is that it allows people to turn off the lights. This may sound obvious, but many of the buildings cited for good daylight design are operated as though daylight did not exist. This is partially the result of inappropriate control systems and lack of user awareness.

Efficient lighting controls can save energy even without daylight

Efficient lighting controls can save energy even without daylight. Lighting systems can be controlled over both space and time to achieve these savings. Some current systems are designed to provide more flexible user control and to prevent energy waste from overdesign. One system also employs photosensors and will be capable of achieving energy savings in daylit offices.

Dimmable and multilevel cut-off systems are only moderately costly

Lighting control options include on/off switching, which has a low to moderate cost but may not realize the full daylight potential and implies uncertain user response. Dimmable and multilevel cut-off systems have a moderate cost, although this is dropping. They add complexity in return for greater savings. Sensing and control devices are available as manual systems, which are fallible but flexible, and as automatic systems, which are reliable but limited.

On/off switching is available on an off-the-shelf basis and has predictable results on fluorescent lamp life. There are potential problems with user acceptance, resulting from sharp changes in lighting levels as one or more fixtures are turned on or off. Experimental results on this issue are mixed. On/off switching can be handled on a circuit-by-circuit basis, fixture-by-fixture, on individual ballasts with a single fixture, or with multilevel ballasts. The latter options, although they involve more expensive switching and control systems, provide effective multilevel responses that may be more acceptable to users.

Dimmable systems are typically more complex and more costly. Although dimmable fluorescent systems are available for specialty applications, there are no widely specified, cost-effective, dimmable fluorescent systems used in the United States.

Dimming need not be continuous. Multilevel step dimming can bridge the gap between on/off switching and conventional dimming systems. A new generation of fluorescent ballasts promises to provide dimming at little incremental cost.

Manual controls are flexible, combining sensitivity and judgment at their best

Given either dimmable or on/off switching, controls can be actuated manually or automatically. Manual controls are flexible, combining sensitivity and judgment at their best. The main danger is that they will be forgotten or unused. Experiments suggest that people in a daylit room will sometimes select higher artificial lighting levels than they will in a windowless room in an apparent attempt to match indoor and outdoor

People will turn lights on more readily than they will turn them off

brightness. People will turn lights on more readily than they will turn them off. But this is a matter of education. Most people prefer the idea of a daylit to an artificially lit space; that preference needs to be supplemented by active participation in the daylit scheme.

Automatic controls must be kept simple to avoid high cost and breakdowns

Automatic controls are more reliable but work without judgment. They must be kept simple to avoid high cost and the possibility of breakdowns. Microprocessors provide virtually unlimited control but represent overkill in a simple office environment.

Selecting controls

Selection of control systems depends on occupancy and function. Perhaps the simplest example is a warehouse with skylights distributed across the roof. Here we can provide simple daylighting over the entire space with a simple control system because the daylight is uniformly distributed. With sidelighting from a window, the daylight gradient into the room becomes significant.

In small offices, work stations should be oriented so that the occupant faces parallel to the window to reduce glare and to provide good contrast and high visibility. With one or two occupants in a small office, there should be little argument over lighting levels, and controls can be simple. Both ambient and task lighting levels can probably be achieved with daylight.

In larger offices, task light may be provided as a supplement to daylight

In larger offices, we find deep bays and open landscape furniture systems. It may be inconvenient to orient all tasks appropriately, and interior spaces may receive insufficient daylight for office tasks. However, daylight may provide adequate ambient light and some design strategies may increase interior daylight performance beyond conventional dimensional limits, which range from 24 to 40 feet. In larger offices, task light may be provided as a supplement to ambient daylight. A relatively simple system can then be used to control an artificial backup for ambient light, while each occupant individually controls task lighting.

Daylighting Codes

Chapter Five

The issue of solar access has become a critical concern as architects and planners start to address the uses of solar energy at community scale. Ralph Knowles has done the pioneer research into the use of a solar envelope as a basis for solar access zoning. This envelope defines the maximum volume of development possible for a given site and set of time constraints without shadowing the neighboring surround. The issue of daylight access in a legal context raises some parallel issues when applied at urban design scale.

The purpose of the daylight code was to establish minimum standards of daylight

J.W. Griffith addressed the issue of daylight in building codes in 1962 when he prepared a model residential ordinance for Chicago. The code was not implemented, but the concept has gained new relevance today. The purpose of the daylight code was to establish minimum standards of daylight for habitable rooms where the Chicago building code already required a window. The scope of the task also included the development of workable techniques for applying and testing compliance with the code, and it included an examination of the impact of daylight controls in a variety of locations and situations. The Griffith report noted that the Chicago building code already required effective clear glass areas of "not less than 10% of a room area." But a window specification is not a daylight specification. So the code further states that the minimum amount of glass should be increased to 25% if the window was obstructed. The code also included dimensional requirements for courts and side yards for daylight purposes.

Griffith's proposed regulation called for 10 foot-candles of daylight in habitable rooms. Since this was a worst-case condition, it was assumed that most rooms would have higher daylight illumination levels. A version of the British daylight factor measurement technique was recommended to determine compliance. Based on an average room depth of 14 feet, Griffith chose a penetration of 7 feet at a height of 30 inches as a measurement point of reference.

In very congested areas, the residential use of lower floor areas could be restricted

Getting more daylight

The implications of the daylight code are of interest today. Sites in desirable locations for daylight could increase density to the limits of existing floor-area ratios and economic parameters. The regulation, however, would encourage orientations, at either the building or apartment unit scale, that would maximize the potential for daylight and minimize obstruction.

Open space requirements would vary with building height and bulk. In very congested areas, the residential use of lower floor areas could be restricted if obstructions prohibited minimal daylight penetration. The relationships between buildings, including new and old construction, would become critical.

Where conformance to the daylight code was based on sky conditions, the maximum density and bulk of a building could be determined by conventional floor-area ratios. Where a reflective component is necessary, the daylight control would fix the height of a building. (Taller buildings may block the light from an obstruction and therefore decrease reflected light.)

The daylight code would not require unusual or undesirable building shapes

The daylight code would affect the orientation of buildings and the dwelling units within them in relation to the surround but would not require unusual or undesirable building shapes.

Daylighting is compatible with many aspects of passive solar design

There are situations where design strategies for daylight would seem to conflict with design strategies for efficient thermal performance. These conflicts, however, are greatest when the designer assumes mechanical solutions to thermal comfort. While daylighting may result in tradeoffs with the version of energy conservation that dictates compact buildings, small surface-to-volume ratios and minimal windows, it is thoroughly compatible with many aspects of passive solar design.

Daylighting implies large window areas that can be oriented and controlled to admit solar heat and light in heating seasons, or only light in cooling seasons. It implies high ceilings, which complement natural ventilation. It implies controls that respond dynamically to the sun. It implies asymmetrical building forms and differentiation by orientation, as does passive solar design.

Daylight also implies an urban design concern for a public right to light that is slightly different but nonetheless consistent with the concept of solar access. And daylight implies a relationship between the user and the natural variations and tempos of day and season, something that has been lacking in most modern urban environments. For many people concerned with the quality of urban life, the bond that daylight can forge between people and natural cycles is its most critical contribution.

Saving energy is just one of the justifications for renewed interest in daylighting

Certainly saving energy is one of the justifications for renewed interest in daylighting. But the energy argument frequently put forward in its defense may not be the correct one. The most frequent argument is that daylight can offset large electrical demands for artificial lighting systems. This is subject to a number of important qualifications. One is that natural lighting cannot reduce energy consumption if the electrical lighting system is not responsive to daylight conditions. Many highly publicized daylit buildings have artificial lighting systems that are designed and operated with no regard for daylight contributions to illumination. Another qualification is that daylight must be considered an integral part of the building design, an important consideration in the location and form of the building. Separating the daylight elements from the overall building and assessing them by a conventional cost/benefit analysis may lead to false conclusions about their cost and energy efficiency.

Daylight must be considered an integral part of building design

Energy has certainly brought new financial support to daylight research and new enthusiasm for the potential of daylight as a design strategy. But energy shortages are not the reason that daylight is important. Far more important are the complement between the natural and artificial environment that daylight facilitates and the improved quality of design it promises. Daylight is a fundamental architectural form issue. Wright, Aalto, Kahn, Corbusier, and other masters knew the functional and aesthetic promise of natural light. We are rediscovering it.

Index